Biodiversity and Environm

An Introducti

This book explores the epistemological and ethical issues at the foundations of environmental philosophy, emphasizing the conservation of biodiversity. Sahotra Sarkar criticizes previous attempts to attribute intrinsic value to nature and defends an anthropocentric postion on biodiversity conservation based on an untraditional concept of transformative value. However, unlike other studies in the field of environmental philosophy, this book is concerned as much with epistemological issues as with environmental ethics. It covers a broad range of topics, including problems of explanation and prediction in traditional ecology and how individual-based models and Geographic Information Systems (GIS) technology are transforming ecology. Introducing a brief history of conservation biology, Sarkar analyzes the new consensus framework for conservation planning through adaptive management. He concludes with a discussion of the future directions for theoretical research in conservation biology and environmental philosophy.

Sahotra Sarkar is Professor of Integrative Biology and of Philosophy at the University of Texas at Austin. He is director of the Biodiversity and Biocultural Conservation Laboratory and works on the design of conservation area networks, primarily in Mexico and India. A former Fellow of the Wissenschaftskolleg zu Berlin, he is the author of *Genetics and Reductionism* and *Molecular Models of Life*.

Biodiversity and Environmental Philosophy

An Introduction

SAHOTRA SARKAR
University of Texas at Austin

CAMBRIDGE UNIVERSITY PRESS
Cambridge, New York, Melbourne, Madrid, Cape Town, Singapore,
São Paulo, Delhi, Dubai, Tokyo

Cambridge University Press
32 Avenue of the Americas, New York, NY 10013-2473, USA

www.cambridge.org
Information on this title: www.cambridge.org/9780521143424

First published 2005
Reprinted 2008
This digitally printed version 2010

A catalog record for this publication is available from the British Library

Library of Congress Cataloging in Publication data

Sarkar, Sahotra.
Biodiversity and environmental philosophy : an introduction / Sahotra Sarkar.
p. cm. – (Cambridge studies in philosophy and biology)
Includes bibliographical references and index.
ISBN 0-521-85132-7 (hardback : alk. paper)
1. Environmental sciences – Philosophy. 2. Bioethics. 3. Biological diversity. I. Title.
II. Series.
GE40.S27 2005
333.95'16 – dc22 2004024334

ISBN 978-0-521-85132-9 Hardback
ISBN 978-0-521-14342-4 Paperback

To Bill Wimsatt, teacher and friend

Contents

Preface

By the mid-1990s, it had become clear that a new interdisciplinary science, conservation biology, was emerging, with concepts, techniques, practices, and traditions of its own and with the explicit goal of conserving biodiversity. The extent to which it was diverging from the disciplines that had spawned it – especially ecology, in which it claimed to have most of its intellectual roots – remained unclear. Two possibilities were clearly present: (i) conservation biology would emerge as a new applied subdiscipline within ecology, one of the many such emerging subdisciplines (for instance, metapopulation ecology and landscape ecology) that were transforming traditional ecology in unprecedented ways, or (ii) it would emerge as a discipline rather distinct from ecology, in part because it was co-opting resources from many other disciplines, including those belonging to the social sciences, and in part because it was explicitly a goal-oriented enterprise with the aim of conserving biodiversity. This normative goal required a type of philosophical justification that is unusual in the customarily purely descriptive scientific context. Conservation biology was both exciting and fashionable in the rich European and neo-European countries (which, along with Japan and a few other rich countries, comprise the so-called North) because it tapped into a growing "environmental movement," which, since the 1960s, had begun to reconfigure the space of the traditional politics of the Left and the Right. Conservation biology was particularly interesting for philosophers of science – and also for anthropologists, historians, social scientists, and others who followed and interpreted the development of science – because it afforded a rather unique opportunity. Philosophers and these other scholars were given a chance to watch a science during its gestation, to see how a new conceptual framework comes to be formulated and refined; how new techniques are introduced, and how they interact with each other and with the developing framework; how technical, social, political, and other constraints modulate the theory and practice of a new science; and, especially, how conventions are introduced on pragmatic grounds and then become reified. It is unfortunate that few philosophers have so far availed themselves of this opportunity.

In the winter of 1996, I taught an upper-level undergraduate course in the Department of Philosophy at McGill University on philosophical issues raised by biodiversity conservation. The explicit agenda of the course was to go beyond traditional environmental ethics and to include discussions of conceptual and epistemological problems at the foundations of conservation biology. There were plenty of conceptual problems: How should we define biodiversity? Can we operationalize the definition? Should operationalization even be a requirement? How much can traditional ecology contribute to conservation biology, given ecology's relative lack of predictive success in the field? How should we make decisions given that we have to produce conservation plans within periods of time far too short for adequate data collection and analysis, let alone for the construction of reliable models? Can we even hope to quantify the uncertainties involved? Where does traditional environmental ethics fit in? What is the role of cultural, economic, and other social factors in conservation biology? What is our place – as individuals committed to biodiversity conservation – in the broader political environmental movement?

This book had its genesis in that course, though it does not claim to broach all of the questions just listed. By the time it was finally written – mainly during the first few months of 2002 – a new tentative consensus framework for conservation biology, dubbed "adaptive management," had finally emerged, though it is as yet impossible to predict how long this consensus will last. This book is ultimately about philosophical problems raised by that framework. Much of this book is exploratory, trying to raise and refine philosophical questions rather than to answer them. Though there is considerable discussion of issues within traditional environmental ethics, the emphasis is on issues that have traditionally occupied the philosophy of science, and on how those issues interact with the normative ethical issues raised by attempts to conserve biodiversity. The material presented here has been used with a fair amount of success in a course called Environmental Philosophy that has been taught in the Departments of Philosophy and Geography and Environment at the University of Texas at Austin for the last six years.

This book is intended as an introduction to the issues that should be discussed under the rubric of environmental philosophy provided that the center of attention is biodiversity conservation (rather than, for instance, pollution, resource depletion, or environmental equity). However, it is not the sort of introductory philosophy textbook that consists of a recitation of the best arguments on both sides of a disputed issue with no attempt to adjudicate between them. (It is not intended as a textbook. At best, it should be regarded as a synthetic introductory text to be used in a classroom only if accompanied by a broad spectrum of materials that disagree with its conclusions.) It attempts to present a coherent anthropocentric position in environmental philosophy that nevertheless emphasizes biodiversity conservation at the possible expense of felt human preferences. It is also highly critical of certain environmental doctrines such as biocentrism and deep ecology. Within the philosophy of science, it defends conservation biology as a discipline radically different from ecology. Probably few, if any, of the analyses presented in this book will survive unscathed as more philosophers turn their attention to these problems – Chapter 8 already expresses doubts about many of them. If this book sparks more work in all of environmental philosophy, beyond just environmental ethics, it will have served its purpose.

Yet another disclaimer is necessary. In attempting to cover a wide range of topics in environmental philosophy, this book draws on a variety of resources beyond traditional philosophical domains, such as environmental ethics and formal epistemology. In the natural sciences, it draws on ecology, evolution, and conservation biology. Beyond that, it draws on topics from economics and operations research. I make no claim to professional expertise in any of these fields, except for parts of conservation biology. Experts in the other areas will probably find grounds for sound complaint against the discussions here – such is the price to be paid for attempting a synthetic work. I welcome all suggested corrections from such experts.

Finally, the "philosophy" in the title of this book is intended to be interpreted in a restricted sense and construed "professionally," that is, as academic philosophy. This choice restricts both the scope of this book and its likely audience. Environmental "philosophy," as popularly understood, often includes a myriad of discourses that have scant resemblance to academic philosophy, from religious ramblings to exhortatory programs designed to arouse indignation against those who would defile Earth. It is not being suggested here that these discourses are devoid of interest or, especially, that such exhortations are not practically and politically valuable insofar as they promote care of our environment. Nor is it being suggested that these discourses are intellectually uninteresting or undeserving of close scrutiny. Nevertheless, they do not constitute philosophy (in the academic sense) and will enter the discussions in this book only peripherally, if at all.

An example will illustrate what is meant here. The fact that some place or some feature of biodiversity may have religious significance or value will sometimes be noted in this book. However, this fact will be interpreted to have only descriptive content and no normative force. The mere fact that some entity has religious value does not, by itself, entail that there is any normative *ethical* reason for us to value that entity. To attribute such a value to that entity would require further argument. In this book, appeal to religious belief will not be taken to provide any foundation for an environmental ethic, no matter how powerful it may be in generating action. It is often powerful. In the Western Ghats of India, a global hot spot of plant endemism and biodiversity, sacred groves are the last remnants of the tropical evergreen wet forests that once covered these mountains. Almost every other place has succumbed to habitat conversion, mainly during the British colonial era. Nevertheless, religious concern does not confer normative value on these important remnants of biodiversity. Some other, *philosophical* argument will be required to attribute that value without denying the pragmatic political utility of these religious concerns. Environmental philosophy will be construed here to fall squarely within the contemporary traditions of professional philosophy.

At the same time, philosophy of science (though not ethics and related areas) will be pursued from a naturalized perspective, as a discipline in continuity with the sciences, dealing more abstractly and self-reflectively with issues that arise within the sciences. There will be no discussion here of questions such as whether biodiversity is a "real property," whatever that might mean, or whether "materialism" or even "naturalism," when those terms are construed metaphysically, can be the foundation for environmental philosophy. Far too much has already been written on such topics without producing any palpable philosophical insight. Answers to questions such as those are also of little relevance to conservation biology. From the perspective adopted here, philosophy of science achieves its validation from the extent to which it can illuminate and, preferably, improve the practice of science itself.

Acknowledgments

For discussions and collaboration over many years, thanks are due to Chris Margules. Many of the themes developed in this book are a result of these interactions, starting at the Wissenschaftskolleg zu Berlin and continuing in the United States and Australia, with partial support from the Commonwealth Scientific and Industrial Research Organization (CSIRO), Australia. Our joint work on philosophical issues connected with biodiversity conservation has been reported in Sarkar and Margules (2002). At the University of Texas at Austin, Camille Parmesan has similarly been a sounding board for many of my ideas on biodiversity conservation and has taught me a lot about the complexities of environmental science. I have also learned much from J. Baird Callicott, Dan Faith, Bryan Norton, and Mike Singer through conversation and other interactions, in addition to their writings.

Several individuals have collaborated with me on many conservation projects over the last five years. From my own laboratory, thanks are due to Anshu Aggarwal, Susan Cameron, Helen Cortes-Burns, Trevon Fuller, Justin Garson, Margaret Heyn, James Justus, Chris Kelley, Michael Mayfield, Kelly McConnell, Jonathan Meiburg, Alexander Moffett, Chris Pappas, Samraat Pawar, Itai Sher, and Ariela Tubert. Besides Margules, other collaborators who have contributed significantly to my thinking about biodiversity and conservation include Jim Dyer, Tony Nicholls, Nick Parker, Víctor Sánchez-Cordero, and Helen Sarakinos. Among those who have helped at different points are Kelley Crews-Meyer, Michael Lewis, Marcy Litvak, Pat Suppes, and Kerrie Wilson. If I have inadvertently missed any individual, I apologize in advance.

For reading the entire manuscript and providing valuable criticisms, thanks are due to Justin Garson, James Justus, Bryan Norton, Ian Nyberg, Jay Odenbaugh, Chris Pappas, Anya Plutynski, Jason Scott Robert, Kristin Shrader-Frechette, Neil Sinhababu, and an anonymous reviewer for Cambridge University Press who sent particularly detailed critical remarks. Others who have provided valuable comments include Trevon Fuller (Chapter 8), Cory Juhl (Chapter 7), Chris Margules (Chapter 5), Alexander Moffett (Chapter 7), and Jessica Pfeifer (Chapters 1 and 2). Thanks are also due to Justin Garson for help with the production of all the figures in Chapters 6 and 7. Finally, this book was completed during a stint at the Max-Planck-Institut für Wissenschaftsgeschichte in Berlin. Thanks are due to that institute and to its director, Hans-Jörg Rheinberger, for support. Much of this book was written at Schwarzes Cafe on Kantstrasse in Berlin, which accounts for its relative lack of coherence.

Biodiversity and Environmental Philosophy

An Introduction

1

Introduction

Some of us have the privilege to live with the type of affluence in which worries about food and shelter never impinge upon our subjective consciousnesses. We succumb to other sources of anxiety. Some, even in the highest echelons of the United States government, are concerned about the possibility of asteroids hitting Earth and the prospects for designing weapons to protect us from them.[1] In recent years, many of us have been worrying about what we call the "environment." Some of us worry about the extinction of species, some about the pollution of our physical surroundings, yet others about changes in our climate. These are all related problems: extinction is often a result of pollution; evidence is mounting that extinction can result from climate change;[2] pollution is a major source of climate change, and so on. The general worry about all these problems constitutes worry about issues that may be broadly categorized as environmental.[3] By and large, explicit worry about most issues in this category, though not about pollution, is confined to the privileged North.[4] The situation is

[1] See <http://impact.arc.nasa.gov/congress>. The gravity of these discussions suggests that they are intended to be taken seriously.

[2] In addition to the obvious plausibility of such a claim (climate changes would make the habitats of many species no longer suitable for them), climate change has been directly, though as yet controversially, implicated in at least one extinction, that of the golden toad (*Bufo periglenes*) that was endemic to the Monteverde cloud forest reserve of Costa Rica (Pounds, Fogden, and Campbell 1999).

[3] It does not matter whether all these problems are lumped together as one worry/concern or kept as separate and distinct worries/concerns. Both usages will be employed here.

[4] For the purpose of this book, the North consists of northern America, Europe, Japan, Australia, New Zealand, and similar developed countries elsewhere. The category is intended to be loosely defined on the basis of economic and cultural features that these countries have in common. The South comprises the underdeveloped countries of the rest of the world. Some Latin American countries and most of the Arab countries do not fall naturally into either category. However, these definitional difficulties are not particularly relevant to the issues that are the focus of this book.

rather different with those for whom everyday life is a struggle for material existence, for instance, for the vast majority of those who live in the so-called Third World (or South) and for many of our own poor or underprivileged. They may never worry explicitly about the general category of the "environment."

This does not necessarily mean that our worries are unjustified or even parochial – a result of an excess of leisure and a lack of social concern for those less fortunate than us. But it does mean that we must be able to justify through adequate argumentation the concerns we express for the environment. This need for justification is especially critical if we believe that these concerns deserve our attention despite equally important, if not more important, concerns we should have for improvements of the material conditions of the lives of the underprivileged. It is easy to condemn the Japanese and the Norwegians for whaling. Neither the Japanese nor the Norwegians are suffering from a deficit of available protein in their diets. Neither face economic collapse if they allow depleted populations of whales to recover to sizes at which they are beyond reasonable danger. In North America, similarly, there is no excuse for continued logging in the Pacific Northwest irrespective of the fate of the northern spotted owl (*Strix occidentalis caurina*). Terminating logging will undoubtedly result in some social disruption. However, these forests are nonrenewable in the short run, and sooner or later logging will have to stop (because, with continued logging, the forests will disappear and there will be nothing left to log). Banning logging completely will almost certainly lead to social dislocation for local communities. However, a rich democratic society should be willing to pay what it takes to ease such social dislocation while protecting a significant part of our dwindling biological heritage, exemplified by the entire forests and not merely by the spotted owl.

But in the South, the situation is not always quite so simple. In 1972, the world's largest *arribada*[5] (arrival for egg laying) of Olive Ridley turtles (*Lepidochelys olivacea*) was discovered on a few beaches in Orissa in eastern India. Once a year, at roughly the same time, hundreds of thousands of Olive Ridleys congregate to form a flotilla a few miles from the coast. This enormous congregation is large enough to be observed from the air. Over the course of a few nights, the turtles clamber in to lay their eggs on the beaches. In one recent instance, between 23 and 27 March 1999, 230,000 females were estimated to have nested in Orissa in one rookery

[5] The term is from the Spanish, because the first instances of this phenomenon were recognized in Costa Rica and Mexico.

alone.[6] Each female turtle always returns to the same beach, which is also believed to be the one on which it was hatched. Between successive returns, it often travels thousands of miles – sea turtle navigation is one of the natural masterpieces produced by evolution. Olive Ridleys are highly endangered and face many threats, including the loss of nesting habitat throughout their range. In 1972, the "Northern worry" that was expressed was a concern over the harvesting of eggs and turtles for food, not only for local consumption but also for export to the markets of Kolkata. Turtle meat was not a delicacy; it was an important source of protein in one of the poorest regions of the world. If our concern for the future of the Olive Ridleys is to trump concern for the livelihood of the poor, we must have very good arguments.

In Orissa, there is cause for optimism. In 1972, while many Northern conservationists clamored for an immediate end to the harvest, others, more responsibly, noted that the need of the poor for food must also be addressed.[7] The Indian government sided with the conservationists but did little to address the concerns of the poor. By 1975, the area had been turned into a wildlife reserve and trade in turtles was banned. However, the Indian state wisely did not enforce the ban to the extent of inducing starvation among the poor. But the presence of the ban probably encouraged the local inhabitants to enlarge their resource base. Eventually, conservation of the turtles found sufficient local support that the conservation measures became successful. However, in the 1980s and 1990s a new threat emerged: death by entanglement in the nets of illegal fishing trawlers. In 1997–98, there were reports of 13,575 Olive Ridley deaths from this source, and the *arribada* had all but disappeared. Proper protection by local voluntary groups, as well as by several government agencies, led to a recovery in 1999.[8] Meanwhile, it has also become far from clear that a complete ban on the harvest of turtle eggs is necessary for successful conservation. Almost all the eggs laid during the first few nights of an *arribada* are destroyed as turtles that arrive later dig nests to lay their own eggs. The eggs that are laid early can be harvested – partly for human consumption, partly for artificial incubation – probably without any harm to the turtle

[6] See Pandav and Kar (2000); the *arribada* has only occasionally been seen since – see the following text.

[7] See Carr (1984), p. 256. To their credit, the conservationists N. Mrosovsky, P. C. H. Pritchard, and H. Harth have all championed the interests of the poor along with the necessity for sea turtle conservation.

[8] See <http://ens.lycos.com/ens/apr99/1999L-04-14-04.html>. After some lean years, the *arribada* was back to its original size in 2004 (Sarkar, personal observations).

population.[9] They would then continue to be an important food resource in a region in which many people continue to eke out a meager existence on an inadequate diet. However, the story of the Orissa *arribada* may be atypical in the sense that conflicts between social justice and biodiversity conservation often have no easy resolution.

Returning to the question with which we started, there are many plausible stories that can be told to justify our concern for the environment. For instance, we may argue that worry about the environment is a long-term worry motivated by a concern for the quality of lives and the state of the planet many generations into the future. Those for whom even the survival of this generation is a matter of uncertainty may not be able to afford such a worry. At least they may not be able to afford it to such an extent that it trumps worries connected to their need for immediate survival. Thus, we should not blame the underprivileged for their lack of explicit environmental concern. At most, we can say that, to the extent that it is consistent with their need to ensure daily survival, even the underprivileged should share our environmental concerns. The poor on the Orissa coast belong in this category. However, those of us who are privileged not only should have such environmental concerns, we should also act on them. Precisely because of our privilege, we have a greater obligation than the less privileged to act in this way. Moreover, such action even indirectly benefits the less privileged: we help to assure the long-term future of their descendants even as they worry about the short-term future through which they must survive in order to produce those descendants. This may well be a good argument, independent of whether it also serves our own privileged interests.

However, to show that the argument of the last paragraph is a good one will require some work. First, it must be shown that the conclusions are logical consequences of the premises. There are two substantive premises explicitly stated in the last paragraph:

(i) worry about the environment is a long-term worry, and
(ii) those for whom immediate survival is uncertain may, at most, be able to afford a long-term worry only to the extent that it is consistent with ensuring their immediate survival,

and two conclusions:

(iii) to the extent that it is consistent with the need to ensure daily survival, even the underprivileged should act on environmental concerns, and

[9] Mrosovsky (2001). The harvesting of early eggs in this way is already successfully practiced in Costa Rica, the locus of another somewhat smaller but better-known *arribada*.

(iv) not only should the privileged have such concerns and act on them, they also, precisely because of their privilege, have more of an obligation to act than the less privileged.

Premises (i) and (ii) are incomplete in the sense that, without being supplemented by other assumptions, they do not logically imply these conclusions. To obtain conclusion (iii), we need at least the following two additional premises:

(v) there is reason to have worries about the environment, and
(vi) we should act on our reasonable worries, long- or short-term.

Further, obtaining conclusion (iv) also requires another premise:

(vii) the privileged have more of an obligation to act than the less privileged.

So far, all that has been shown is that the conclusions finally follow from the assumed premises, that is, that it is not possible for the premises to be true and the conclusions false.[10] But notice that this does not tell us much: it does not tell us that the conclusions are, in fact, true and that (therefore) we should believe them.[11] Crucially, we have to establish claim (v), as well as another assumption that is deceptively embedded in premise (vi):[12]

(viii) these worries are of the sort that require action on our part.

Without assuming claim (viii), we have no obligation to respond to our worries by intervening in the world; we do not even know that our worries are more important than, say, a worry about the heat death of Sun. Not all worries deserve action; for some, such as the worries of hypochondriacs, the desirable response is precisely the opposite. Are biodiversity conservationists more like doctors than hypochondriacs? "Environmental skeptics" do not believe that the answer is obvious;[13] if we genuinely believe that our environmental concerns are justified, we must be able to answer such skeptics.

Establishing claim (v) requires a close look at the state of the world. Most importantly, it requires judging what we should – and should not – believe about the world given the empirical data we can collect, the best scientific

[10] In philosophical terminology, the argument is *valid*.

[11] In philosophical terminology, if the premises are true in a valid argument, the argument is then *sound*.

[12] The other premises are being taken as unproblematic.

[13] See, for example, Lomborg (2001), who doubts that extinctions are occurring at a rate significant enough for there to be reason to worry. See also Chapter 5, § 5.3.

models we have, and the uncertainties involved in both data collection and model construction. Ecology, the science to which biodiversity conservationists most routinely turn for empirical support, is notorious for being unable to make claims about the world with much certainty. For instance, are species' extinctions taking place at a much higher rate than the normal rate (that is, the average or background extinction rate during evolutionary periods other than mass extinctions)? Those who believe that the rate is several times higher than normal often talk of a biodiversity or extinction "crisis." Crisis talk has tremendous rhetorical value in the political terrain of the North. Whether it is justified is entirely another matter. Moving on to another topic, are we certain that climate change is taking place? Crucially, are we certain that these processes, if they are occurring, are occurring for anthropogenic reasons (human interference with the natural world)? If they are not, merely modifying some aspects of our usual behavior may not solve the problems. Answering these questions involves normative considerations that are epistemological: what *should* we *believe* (the *should* reflecting normativity and the question of *belief* placing these considerations centrally within epistemology). We must try to answer these questions with such intellectual rigor as to silence skeptics (provided that they also accept such a requirement of rigor).

Establishing claim (viii) also involves normativity, but this time the considerations are ethical and, on occasion, aesthetic. Even if it is taken as established that there are serious ongoing environmental problems, it does not immediately follow that we have an obligation to act. To get to such an obligation, we have to analyze carefully the nature of our relation to the environment. For instance, we have to assess whether the environment embodies some sort of value that requires us to refrain from harming it, in the same way that we have an ethical obligation not to harm another human being. Or perhaps, alternatively, we should not harm the environment because of how we value humans themselves. For both alternatives, moreover, it may be the case that we have not only an obligation not to harm, but also a positive obligation to nurture features of our environment (for instance, endangered species at risk of impending extinction).

Suppose that we have successfully established claim (viii). Epistemological considerations return to haunt us when we take seriously the obligation to act and begin devising strategies to protect the environment. Moreover, now they often become intertwined with ethical and aesthetic ones. Some of the purely epistemological considerations are similar to those mentioned earlier. Given our empirical knowledge, our models, and all the uncertainties involved, what policies are likely to produce the desired results? There also is a radically new element. Environmental change is often irreversible: recall the much-used slogan "Extinction Is Forever." Not only must we act under uncertainty, but sometimes we may also feel that it is imperative to act immediately. We may not have the time to collect all the data we would like or to satisfactorily complete theoretical analyses before acting. Suppose, for instance, oil interests threaten a wildlife refuge.[14] Oil industry managers claim that oil exploration will have, at most, a minimal negative impact on the future of wildlife. The data needed to establish the veracity of this prediction would take years to collect. But in the political arena, the oil industry must be answered now. We have to decide whether we are willing to accept the risks involved in oil exploration. If we opt to allow exploration and the industry managers' predictions about the impact on wildlife are wrong, the harm done to wildlife may be irreversible. Somehow, we must navigate between our epistemological uncertainties and the potential ethical, aesthetic, and even economic costs of irreversible harm to wildlife and other components of biodiversity. Scenarios such as these are not at all uncommon: the fossil fuel industry threatens habitats almost everywhere in the world; logging threatens rainforests; the Japanese and Norwegians seem to have an irreplaceable appetite for whale meat; elephant ivory from Africa remains a lucrative trade.

1.1. A FOCUS ON BIODIVERSITY

These are the sorts of issues that concern environmental philosophy: like the rest of philosophy, it is inherently a normative discipline. This book will explore how environmental philosophers and scientists (and others) analyze and navigate these issues when confronted with concrete environmental problems in everyday practice, but with an emphasis on general conceptual issues and the principles on which policies are based rather than on contextually contingent detail. The focus of this book will be on the loss of biodiversity, which will be taken to be exemplary of the environmental problems we currently face. ("Biodiversity" will be formally defined in Chapter 6, § 6.4; devising a satisfactory definition is not easy. For now, it will be used informally to refer to diversity at all levels of taxonomic organization, particularly the level of species.)

[14] In the United States in 2004, this is not a purely hypothetical example: oil interests, with the active support of an erstwhile oilman president, continue to threaten the pristine Arctic National Wildlife Refuge.

However, except for one chapter (Chapter 6), which is specifically concerned with the details of conservation biology as it is currently practiced, virtually all of the arguments and analyses will be applicable to many other environmental contexts. Issues such as the difficulty of estimating parameters and teasing out definite predictions from ecological models, coping with unquantifiable uncertainty, and yet having to make decisions with irreversible consequences, permeate all thoughtful discussions of biodiversity conservation. But they are also equally relevant to all other significant environmental concerns. For instance, they have been central to the recent debates about the existence and anthropogenic etiology of climate change, and about the risk posed to communities by the storage of nuclear or other hazardous wastes in their neighborhoods.[15] Thus a focus on biodiversity conservation does not illegitimately constrict the issues considered in environmental philosophy. Except for the fact that the present context requires the use of ecological models, many of these epistemological discussions – in particular, the analysis of uncertainty – are also relevant to many other contexts in which science interacts with society, for instance, medical contexts.

At least to some mitigated extent, the focus on biodiversity can also be defended by arguing that many other, though certainly not all, environmental concerns can ultimately be accommodated within an exploration of our concern for the persistence of life on Earth. It is worth examining how far this argument can be pushed. To the extent that it is legitimate to distinguish between biological and cultural requirements for human life and well-being, these other environmental problems are generally perceived as such because they mainly threaten the *biological* basis of human life.[16] For instance, pollution biologically threatens human health and life in the short term; climate change is similarly threatening in the long term. But pollution also threatens many other forms of life wherever it occurs: oil spills from tankers are notorious for devastating marine life. Climate change has already been implicated, though not yet uncontroversially, in at least one extinction, that of the remarkable golden toad, *Bufo periglenes*, in the otherwise pristine Monteverde cloud forest reserve in Costa Rica.[17] These are typical cases.

[15] See, for instance, Shrader-Frechette (1991), a work that is primarily concerned with pollution. Much of that discussion of risk can be co-opted without modification for use in the context of biodiversity conservation.

[16] This, perhaps, helps to explain why environmental politics often does not respect the traditional *cultural* political boundaries between the Left and the Right.

[17] See Pounds, Fogden, and Campbell (1999). The extinction of the golden toad is usually taken to be an exemplar of a general amphibian decline "crisis" – see Sarkar (1996) for a critical discussion of the issues raised by this and similar examples.

Any biological factor that is systematically threatening human life will also threaten many other forms, at least of animal life: human biology is not special at this level of generality. Consequently, exploring the problems that threaten life on Earth in general – that is, biodiversity – ipso facto includes an exploration of these other problems. If anything, what is special about biodiversity depletion is that it also raises other environmental problems, for instance, concerns about seemingly useless pieces of land, desert habitats, remnants of prairie in abandoned rail tracks in the North American Midwest, and so on.

While biodiversity depletion should perhaps be central to any exploration of environmental philosophy, it would be illegitimate to suggest that the issues covered in this book exhaust the scope of environmental philosophy. For instance, many political issues, including those concerning equity and environmental justice that arise in the context of discussions of pollution, and that are central to political philosophy, are usually not discussed in the context of biodiversity depletion.[18] But even here, it is difficult to draw a clear line. The claims made at the end of the last section implicitly assume that there is a higher social cost involved in compromising the diet of the poor than in changing the diet of the rich. This is an assumption about equity, and it arose in the context of biodiversity conservation.

Finally, it is probably worth noting that the problem of biodiversity conservation does not exhaust all the themes that fall within the practice of contemporary conservation biology. Two themes that will not be explored here are those of sustainability and restoration.[19] Both merit much more philosophical attention than has so far been afforded to them. This book remains limited in its scope even within the domain of philosophical issues raised by conservation biology.

1.2. THE STRUCTURE OF THE BOOK

The remaining chapters of this book will explore a variety of issues connected with thinking about the environment, particularly biodiversity, and the interrelations among these issues. The purpose of this book – as noted in the Preface – is not to defend any single position definitively. Rather,

[18] For a recent philosophical exploration of issues surrounding environmental justice, see Shrader-Frechette (2002). Important earlier works include, especially, Wenz (1988).

[19] In the case of sustainability, Norton's (2003) collection of essays, *Searching for Sustainability*, provides a start.

it is to lay out a set of related tentative positions that deserve further exploration. Throughout, because of the potentially broad interest in environmental philosophy, an attempt has been made to avoid unnecessary philosophical jargon, whether it be in ethics or epistemology. Roughly, the first half of the book (Chapters 2–4) concerns ethical (and, to a lesser extent, aesthetic) rationales for biodiversity conservation (and, somewhat more generally, other aspects of the conservation of nature). What is defended is a broadly anthropocentric perspective for biodiversity conservation. The second half of the book (Chapters 5–7) explores epistemological issues. "Epistemology" is being construed here broadly, to include questions about the so-called logic of the sciences (the confirmation of scientific models and theories, the quantification of uncertainty, etc.), the representation of nature in models (realism versus instrumentalism, etc.), as well as the relationships among the various sciences. The most important claim defended in the second section of the book is that conservation biology should be viewed as a discipline distinct from ecology. The structure of conservation biology is also explored. Together, these two parts provide a reasonably complete exploration of the philosophical issues arising from biodiversity conservation.

It is important for the program pursued in this book that the normative justification for biodiversity conservation be developed first, before the discussion turns to the technical foundations of conservation biology. The particular justifications for biodiversity conservation that are found acceptable will constrain what is permissible in the practice of conservation, for instance, whether an accommodation of identifiable human interests alone is sufficient for the design of an ethically defensible conservation policy. The justification for biodiversity conservation defended in the first part of this book is anthropocentric. Consequently, the framework developed in the second part for incorporating both biodiversity-related and other human values into a conservation plan employs techniques co-opted from the social sciences. Had the first part of the book concluded that biodiversity conservation has its ethical basis in nonhuman values and interests, no such seamless transition between biodiversity-related values and other human values would have been possible.

Chapter 2 consists of a philosophical exploration of the nature of our concern for the environment at a very general level. It explores the question of what it means for us to be concerned about the environment. In particular, it explores the role played by two myths, that of lost futures and that of the golden age, that often lie behind this concern. The latter myth is also shown to lie behind the concern for wilderness preservation, which is a dominant concern in many of today's affluent societies. Wilderness preservation is compared to and contrasted with biodiversity conservation. This chapter introduces many themes and arguments – for instance, the potential human benefits of biodiversity – that are treated more fully in later chapters. The main focus in this chapter is on the sources of our environmental concerns, and not on their justification. But it is equally concerned with giving at least a pedestrian introduction to the intellectual background from which environmental philosophy has recently emerged. However, Chapter 2 is not strictly necessary for the conceptual development of the program of this book, and readers may choose to skip it (along with the final section of the present chapter).

Having a concern does not imply that the concern is justified. Chapters 3 and 4 turn to the question of justification that has traditionally been at the core of environmental ethics (though, as we shall see, aesthetics is also central to the type of normative considerations that become relevant): why we should conserve biodiversity. The treatment of environmental ethics is not intended to be comprehensive – any such attempt would go well beyond the confines of one part of one book. Instead, the treatment is intended to provide just enough detail to constitute an adequate background for the epistemological discussions of the second half of the book. In Chapter 3, adequacy conditions for an ethic of biodiversity conservation are introduced. The discussion then moves to the question of whether biodiversity (in particular, any nonhuman species) has intrinsic value and the relation of such an attribution to a position variously called "biocentrism" and "ecocentrism." Attempts to ground a conservationist ethic on such attributions of intrinsic value are found wanting in spite of their popularity among conservation biologists, at least in North America, and among some environmental ethicists. Chapter 4 considers the alternative that ethical considerations about biodiversity are ultimately based on human values and interests. However, these human values cannot all be reduced to simple demand values determined in the marketplace. Some of them get their value because of their ability to transform demand values: they thus have "transformative value." An associated position, called "tempered anthropocentrism," establishes some of the connections between environmental values and aesthetic appreciation.

Moving beyond the justificatory project, from Chapter 5 onward, epistemological issues emerge as the center of attention. Because ecology is supposed to be the scientific discipline on which attempts to conserve biodiversity must be founded, Chapter 5 provides a very condensed critical analysis of the science of ecology to the extent that it contributes to biodiversity conservation. The emphasis is on the difficulty of using traditional ecological models in the applied context of biodiversity conservation. Some recent developments that give grounds for cautious optimism (Geographic Information Systems [GIS]–based and individual-based models) are also briefly discussed. Finally, Chapter 5 ends with a short discussion of what we truly know about extinction, distinguishing that knowledge from what may be only inflated rhetoric. The current rate of extinction does turn out to be a serious problem that has to be faced. (Thus it provides support for premise [vii] of the argument with which this chapter started.) However, it may not justify talk of an extinction "crisis." Much of this discussion traditionally belongs to the philosophy of science, which, unfortunately, has usually been ignored in environmental philosophy.

Chapter 6 analyzes the structure of contemporary conservation biology, which primarily consists of a consensus framework for adaptive management that has emerged during the last decade and that has become central to biodiversity conservation planning as it is practiced today. The differences between ecology and conservation biology are implicitly emphasized throughout this chapter. The consensus framework consists of a set of practices that can be said to comprise doing conservation biology. These practices are both empirical and theoretical. In the interest of initiating widespread philosophical analysis, the theoretical practices, that is, the problems that have to be solved, are distinguished from the empirical ones and receive special attention. These are (i) the place prioritization problem, (ii) the surrogacy problem, (iii) the viability problem, and (iv) the feasibility problem. Discussion of the last problem explicates how normative (sociopolitical and ethical) considerations are incorporated into the consensus framework. Chapter 7 takes up the problem of synchronizing incommensurable objectives – for instance, biodiversity conservation, economic cost, and social cost. It tries to show how far synchronization can be achieved using objective standards, that is, without imposing ad hoc measures for the purpose of spurious quantification. Finally, it grapples with the difficult problem of coping with uncertainty, at times when decisions must be made in a limited time with inadequate data and models.

The last chapter, Chapter 8, is not a conventional conclusion. Rather, it points out areas where much more philosophical work needs to be done beyond the discussions of this book. New issues and concerns are introduced, both to refine, modify, and move beyond the analyses of this book, and to encourage exploration of issues in environmental philosophy that are either ignored or receive short shrift here.

1.3. THREE FLAWED ARGUMENTS

Philosophical reflection typically takes us into uncomfortable territory – this may be unfortunate, but that is the nature of the discipline. Take abortion. Conflicting prejudices, generally of a political or religious nature, may have us believe either that it is murder or that it amounts to little more than a visit to a doctor to reduce a minor irritation. Philosophical analysis provides the disputants with no such easy way out. The beginning of human life is hard to define, and abortion is not a simple moral problem: we can have little philosophical (as opposed to pragmatic) justification for whatever position we adopt. Or take cannibalism. For members of most contemporary societies, it is repugnant. But it is hard – though probably not impossible – to find philosophically justifiable reasons to prohibit the eating of the flesh of a human who died of natural causes other than infectious disease. Or take human cloning. As many philosophers have recently pointed out, the philosophical case against human cloning is hard to substantiate, though many people find the new possibility intrinsically repugnant.[20]

Environmental issues are no different in this respect. This chapter will conclude by analyzing three arguments that are dear to many environmentalists but are of little substantive, rather than rhetorical, value. Though all three arguments are faulty, they get progressively better, and the last two are not entirely unreliable. Moreover, as will be explicitly pointed out, each of these arguments contains an important kernel of insight; unfortunately, in each case, it gets shrouded in the rhetoric. The reason for this critical exercise is to illustrate a certain style of analysis that this book aspires to follow: a demand for rigor that will require following assumptions to their logical conclusions even when these are psychologically or politically unsettling. This book will attempt to live up to that standard because philosophy is important not just when it exposes what we truly have good reason to believe; it is equally, if not more, important when it shows that we have no such reason to accept much of what we commonly believe. However, this section contains no material that will be needed to follow the general program of this book, and it can be safely ignored by those readers who wish to turn immediately to the positive program.

The first of the three flawed arguments is the "rivet argument," often invoked to justify measures designed to protect every species from

[20] See Pence (1998) for an interesting defense of human cloning; Kass and Wilson (1998) provide a counterpoint.

extinction.[21] Planet Earth is supposed to be similar to an airplane. Given an airplane in reasonably good condition, we can conclude that the loss of a single rivet will not make the plane unsafe. But if we allow that rivet to be lost, the argument goes, we will slither down a slippery slope. Sooner or later, the next rivet will be like the proverbial last straw that breaks the camel's back. The plane will be doomed when we lose that rivet. Each species on Earth is supposed to be like one of these rivets. We may allow one to go to extinction and continue to live with no real fear of the entire biosphere collapsing. But ultimately, as more species disappear, it will collapse. Just as we do not really know which rivet is the last, we do not know which species will mark the end of our biological world.

Most biodiversity conservationists will probably admit that this argument has strong rhetorical value, especially because it attempts to justify the conservation of every species, not just those for which we can specify a definite positive value at present. Nevertheless, it is a flawed argument. For one thing, it is an argument only by analogy, no stronger than the strength of the analogy. Is the biosphere really that much like an airplane? Or, to borrow an analogy from Hume, is it perhaps more like a vegetable?[22] For most vegetables, the loss of a leaf (or of many other parts) will lead only to the growth of another one. Thus, under the vegetable metaphor, there is no reason to suppose that the biosphere will collapse after another extinction. Moreover, there are some, though not very good, reasons to suppose that the vegetable metaphor captures something interesting about evolutionary processes: an extinction may well free up space for new speciation. After all, almost every mass extinction in the past has probably been followed by a burst of speciation renewing the diversity of life on Earth. However, the anthropogenic extinctions of today are probably unlike all past mass extinctions: they may well be altering the surface of Earth in a way that makes it impossible for any living organism to survive. Thus the last point should not be taken to suggest, even if we largely accept the vegetable analogy, that we should have no concern for the ongoing extinctions.

Even without appealing to the general failure of arguments by analogy, there is reason to deny the force of the rivet argument. Loss of all of the

[21] See Ehrlich and Ehrlich (1981), pp. xi–xiv, for the original argument. Takacs (1996), pp. 53–54, analyzes its important rhetorical role in attempting to justify the conservation of *every* species, even when the individual value of a particular species is not known.

[22] Hume (1948 [first published in 1779]) brilliantly deploys this analogy to argue against the design argument for the existence of god, which uses an analogy between Earth and a complicated machine. Will those conservation biologists who believe in the strength of the rivet argument also buy the design argument?

rivets in a seat inside the airplane will lead to no major disaster. We can let a lot of these disappear, saving our concern for those rivets that are known to make a functional difference to the safety of the plane.[23] Loss of a species of bacteria, or even a large predator, will not lead to the collapse of an ecological community (a group of interacting species sharing the same habitat – see Chapter 5, § 5.1). Tigers (*Panthera tigris*) have disappeared from most of their original range during the last century. To the extent that we can tell after so short a period of time, though there are good reasons to mourn these disappearances, the biological communities in these habitats have not themselves disappeared. Although made poorer and less interesting by the loss of tigers, these communities have nevertheless persisted. At most, what can be said is that, if there are so-called keystone species that, by definition, are critical to the continued persistence of a community, then these species deserve special attention. But even this does not provide an argument for the preservation of every species. It only provides an argument for the preservation of keystone species. It may now be argued that we do not know which species are keystone species. That is often the case, but even then, as our knowledge grows, we will systematically lose any rationale for preserving more and more species because they will be known not to be keystone species. This is no argument for general biodiversity conservation. It should also not be forgotten that, as the years have gone by, it has become empirically less and less clear that most ecological communities have keystone species.[24]

The kernel of insight in this argument is that there may be many cases in which we do not know how critical a particular species is to the survival of a community. In the absence of any other knowledge, because of the irreversibility of most ecological changes, it would be imprudent to assume that a particular species does not matter. Beyond this kernel, moreover, there is another version of this argument, one that is not quite a rivet argument but that is logically sound and much more important. Lovejoy puts it thus:

> The loss of a single species out of the millions that exist seems of so little consequence. The problem is a classic one in philosophy; increments seem so negligible, yet in aggregate they are highly significant. . . . But when the increments are in singletons, tens, or even thousands of species out of millions, such effects may be imperceptible, and may seem even more so when many of the

[23] More generally, all such slippery slope arguments are faulty for this reason. See Pence (1998), pp. 144–146, for an elaboration of this point.

[24] See Power et al. (1996); however, keystone species may exist and yet be unrecognized because of the empirical difficulty of demonstrating their role in a community.

> effects are delayed or are impossible to measure. . . . By the time the accumulated effects of many such incremental decisions are perceived, an overshoot problem is at hand.[25]

Lovejoy's argument differs from the rivet argument insofar as it does not suggest that the extinction of a *single* critical species may suddenly lead to the collapse of an ecosystem. Rather, he argues that the accumulated effects of an attitude of permissiveness toward continuing individual extinctions may be much larger than is usually appreciated. The conclusion is that it is wise policy to assume, as a precautionary principle, that every species matters. The arguments of this book will endorse that conclusion.

The second flawed argument that is sometimes invoked in the context of biodiversity conservation is a version of the "tragedy of the commons" argument.[26] Suppose that biodiversity is part of the common property of humanity. If each human individual (or group) utilizes more than a fair share of that property, then that individual (or group) accrues a personal benefit at no cost. Thus all individuals (or groups) have an incentive to take more than the fair share. The result is a total destruction of the commons. There is nothing wrong with the argument so far; it is certainly valid. The trouble begins when it is then concluded that individuals (or groups) cannot be left to make decisions about the use of the commons for themselves. Some versions of the argument go on to call for draconian curtailment of the uses of the commons that may be enjoyed by individuals (or groups). Garrett Hardin, for instance, urges "the necessity of abandoning the commons in breeding. . . . The only way we can preserve and nurture other and more precious freedoms is by relinquishing the freedom to breed, and that very soon."[27] The "other freedoms" are those that are supposed to make human life worthwhile. For Hardin, state control is the only way in which a ruinous growth of the human population can be controlled. In order to reach such stronger conclusions, it must be assumed that the argument is not only valid but also sound. But, historically, at least one of the premises is known to be false in typical cases: that individuals (or groups) necessarily benefit by taking more than the fair share. Community values often act to ensure that there is a cost to taking more than the fair share. Local control of resources often does not lead to the depletion of biodiversity or other common goods. For instance, the Sàmi reindeer herders of northern Norway divide and

[25] Lovejoy (1986), p. 22. Norton (1987), p. 67, calls this problem a "zero-infinity" dilemma.

[26] For the original argument, see Hardin (1968). The objections raised in the text are just as applicable to Hardin's original argument as they are to the modified version discussed here.

[27] Hardin (1968), p. 1248. For a different critical response to Hardin, see Cox (1985).

regroup their herds several times a year so that common pastures are not overgrazed. Social castes in many areas of India have come to divide biotic resources in such a way that each group uses only part of the resources potentially available to it and no resource is threatened. There are *many* such examples.[28]

One can even argue that, from the perspective of human cultural evolution, the emergence of such practices should come as no surprise. Groups of individuals (or of smaller groups) survive only if they have learned to manage the commons in a sustainable fashion. Otherwise, they would have become culturally extinct. Rather than decree global measures on the use of biodiversity (and other resources), conservation biologists could learn much from the local management of the commons. Nevertheless, in order for this to be likely, there are two critical assumptions that must be made:

(i) that the groups have not grown in size to such an extent that practices that were once sustainable can no longer be sustained, and
(ii) that the relevant cultural practices have not changed in such a way that they no longer guarantee the continued survival of the group in its environment.

It is easy enough to see why both these assumptions are suspect. Many groups are now growing at much higher rates than in the past, due in part to decreased childhood mortality as medical care has improved. Moreover, technological change has reached almost every corner of Earth. Forests that survived handheld saws have little chance against mechanized logging. Sociopolitical changes have often led to the erosion of traditional community values and authority. The tragedy of the commons may not be universal. But neither is it nonexistent. What is perhaps most striking is that the obvious inference to be drawn from the argument to the tragedy of the commons is that unconstrained liberal individualism may be the source of the tragedy: one way or another, controls over individual behavior in consumption are necessary. But this inference goes against the spirit of liberal capitalism, and Hardin does not even consider it. It is not surprising, given that the political ideology of liberal capitalism dominates our era, that the conclusion that suggests a radical political transformation of our society is the one that is avoided. Nevertheless, the implied criticism of liberal individualism constitutes the kernel of genuine insight in this argument.

[28] See Bjørklund (1990) for the Norwegian case, and Gadgil and Berkes (1991) for the case of India as well as several other examples.

The third flawed argument is the "population explosion" argument.[29] Human population growth is taken to be the reason why biodiversity depletion (especially in the form of the extinction of species) has accelerated in the recent past. It is concluded that controlling population growth would help to halt, or at least to slow, this depletion. So far, the argument is not only valid but also sound – this is its important kernel of insight. But now, in most deployments of this argument, progressively stronger conclusions are drawn that monotonically decrease in merit. The first such extrapolation is to conclude that consciously and systematically controlling population growth *everywhere* is *necessary* to prevent a continued accelerated rate of biodiversity depletion. What is forgotten during this extrapolation is that what is critical is not the size of a population but the resources that it consumes, for instance, the size and significance of the habitats it modifies, and the nature of that modification.[30] The total consumption of a population, which is probably a reasonable measure of its negative impact on biodiversity and many other environmental parameters, is the product of the population size and the amount that each individual consumes. If a small population, such as that of Japan, systematically logs the rainforests in Borneo only to use the timber for trivial purposes and without recycling,[31] a constant population of Japan will nevertheless lead to the depletion of all of the biodiversity that is endemic to Borneo. A systematic growth in the population of the indigenous ethnic groups of Borneo, provided that it is not too large, may have no significant effect on the biodiversity of Borneo. Only empirical research can determine what constitutes the range of permissible population growth. Picking population growth as the only relevant etiological factor, or even as the most salient one, without such an empirical investigation is illegitimate. Moreover, there is no evidence to suggest that curbing population growth is any easier than transforming consumption practices. Thus there may not even be pragmatic rhetorical value to this elaboration of the original argument.

But there is worse to come. An even stronger extrapolation would equate the potential negative impact of the population growth of some nonindustrialized country, such as Bangladesh, with that of an industrialized one, such as Canada. But the patterns of consumption are radically different.

[29] Ehrlich and Ehrlich (1968) were probably the first to emphasize this argument. However, concern for human population growth had already been emphasized by Boulding (1966) – see Chapter 2, § 2.1.

[30] See Cohen (1995) for a detailed elaboration of this point. Cohen's book remains the definitive treatment of the question as to how many individuals Earth can support.

[31] For more on this example, see Bevis (1995).

Though it is hard to defend exact numerical values – estimates range from ten to thirty – the rate at which resources are consumed on the average by a Canadian resident is several times higher than the rate at which they are consumed by a resident of Bangladesh.[32] Consequently, on a global scale, a much higher population growth rate in Bangladesh is probably statistically irrelevant to the environment compared to a much more modest rate of growth in the Canadian population. The global scale is the relevant one for many environmental problems, such as climate change through global warming and other factors. Yet there is often more worry about the admittedly high (though decreasing) population growth rate of Bangladesh than about the modest growth rate in Canada. Some European countries, such as Germany, even encourage population growth by means other than immigration: they subsidize the production of indigenous offspring. While perhaps no one has explicitly equated the per capita environmental effects of population growth in the South and the North, almost any policy prescription that lumps together populations from the South and the North to suggest uniform quantitative requirements implicitly involves this extrapolation. This includes the original form of the Kyoto protocol, which had no special dispensation for the countries of the South.[33]

An even stranger extrapolation of the population explosion argument, and one with worrisome political overtones, has been to use it to argue for a limit to immigration to affluent societies from those that are not.[34] A "local" reason for this conclusion is the general population explosion argument: it is desirable to control population growth in these affluent societies as much as possible, just as it is desirable to control population growth in general. An additional "global" reason, also sometimes given in defense of this conclusion, is partly based on the arguments endorsed in the last paragraph. The claim is that potential immigrants to Northern societies would adopt

[32] See, for instance, Guha (1989). Athanasiou and Baer (2003), p. 22, graph the annual per capita carbon dioxide emission, which is a measure of energy consumption, for both Bangladesh and the United States. It differs by a factor of at least fifty.

[33] Subsequent modifications did take these inequities into account – see Athanasiou and Baer (2003).

[34] In the United States, a dispute over this issue led to serious divisions in the Sierra Club in the late 1990s. The political undertones of this debate are not hard to detect, though they have not been analyzed in sufficient detail – see van Gelder and Rysavy (1998). The arguments mentioned in this paragraph of the text were all deployed during the Sierra Club debate. Ultimately, the Sierra Club voted to remain neutral on immigration policy. However, the debate resurfaced in 2003 – for an analysis, see the report published online by the Center for New Community <www.newcomm.org>. See also Hardin (1995) for arguments in favor of limits to immigration on alleged environmental grounds.

overconsumptive habits, whereas if they remained in their societies of origin, they would consume much less. It is hard not to suspect that the etiological origin of this argument is often more politically motivated than it is based on concern for the environment. For instance, the following scenario is never mentioned: political activism could be used to reduce overconsumption in the affluent societies to such an extent that consumption would fall well below sustainable levels. It is reasonable to assume that this could be achieved without making human lives unbearable or even particularly uncomfortable. Then further immigration from overpopulated, less affluent societies could be encouraged. This would have the added benefit of decreasing the total consumption in those societies, which is in some cases also not sustainable, but which cannot be decreased without reducing human lives to conditions unthinkable in affluent societies. The purveyors of population explosion arguments seem strangely silent about such political alternatives.

Making flawed arguments unknowingly marks poor philosophy; making them knowingly marks a loss of intellectual integrity. These are good enough reasons never to use such arguments, even for rhetorical purposes. However, and this is an important point, there is also another compelling prudential argument for never making poor arguments in contexts such as that of biodiversity conservation. As later chapters of this book – particularly Chapters 6–8 – will note, biodiversity conservation occurs in contested political terrain. The management of land for the sake of biodiversity conservation is only one potential use of that land, and it has to compete with other potential uses that often require the conversion of living habitats into abiotic patches, or at least into systems that are less biologically diverse. Sound short-term economic interests may well argue for such other uses, for instance, the creation of industrial complexes or monoculture plantations. Conservationists and their opponents have to present their cases to the public, either to the entire polity or to its elected (or otherwise appointed) representatives, who will then decide the future of the land. Not only are poor arguments obviously vulnerable to refutation by opponents, but their purveyance may lead to an increasing distrust of conservationists by decision makers. Such a potential loss of credibility should be regarded as a serious issue for conservationists. Recourse to poor argument is one way in which biodiversity conservationists may lose credibility, but it is not the only one. Inflated rhetoric – for instance, the potentially unjustified reference to a "crisis" when talking of the current extinction rate – is another way in which biodiversity conservationists may lose public credibility. (This issue is further discussed in Chapter 5, § 5.3.)

2

Concern for the Environment

Why are arguments such as the rivet argument, the alleged tragedy of the commons, and the population explosion argument so compelling for so many, despite their faulty premises and, in the case of the rivet argument, faulty logic? At a superficial level, it is simply because many of us want to believe the most dire implication of their conclusions: that there is something radically wrong with our present relationship with the world around us, that is, in the widest sense, with our "environment." For those who do not have to worry about food today and shelter tomorrow, this sometimes results in a long-term worry about the future. However, once we enter the context of philosophical analysis, our concern for the future deserves a deeper scrutiny. Is our worry appropriate? This will be an underlying, though often only implicit, concern of much of this book, particularly Chapters 3 and 4. Equally importantly, how and why does such a worry arise? This is the problem broached in this chapter, though no definitive answer will even be attempted. It will be suggested that two deeply influential myths, which impinge upon our conscious ideologies (broad sociopolitical normative frameworks for interpreting the world),[1] modulate the nature of our concern for the environment.

The discussions of this chapter are not strictly necessary in order to follow the development of the main theme of this book as outlined in Chapter 1 (§ 1.2). However, the two myths discussed here, those of "lost futures" and the "golden age," help to situate the systematic positions of later chapters

[1] The term "ideology" is used in this book only in the loose sense indicated here – for more on the controversy over what "ideology" may be presumed to mean, see, for example, McLellan (1986) and Eagleton (1991, 1994). "Ideology" is not intended necessarily to have a negative connotation. The most important feature of "ideology," as used here, is that an ideology can be fully articulated as a set of explicit beliefs. Many uses of "ideology" do not satisfy this requirement. "Myth" will be defined later (in § 2.1).

in the intellectual – and in part sociopolitical – contexts from which they emerged. Both myths will be carefully explicated later. Suffice it to note here that the myth of lost futures (§ 2.1), a vision of future environments bereft of both intellectual and practical resources, is supposed to provide the background that makes the anthropocentric position outlined in Chapter 4 appear compelling and, therefore, worthy of philosophical defense. To the extent that all subsequent chapters are based on that position, this myth remains at the background of much of the discussion of this book. Similarly, the myth of the golden age (§ 2.2), a vision of a fall from grace, lies at the background not only of wilderness preservationism (§ 2.3) but also, though only to a limited extent, of some of the positions criticized in Chapter 3.

From one point of view, concern for the environment is almost as old as recorded history; from another, equally extreme point of view, it is of very recent vintage and began only in the 1960s. Those who adhere to the former perspective will point to examples such as the Mauryan emperor Asoka (299–237 B.C.E.) of India, who ordered the preservation of forests crucial for maintaining the supply of elephants (*Elephas maximus*) for the royal army.[2] Those who adhere to the latter will often invoke the first Earth Day (22 April 1970) as their foundational myth.[3] For many, a critical event was the publication of Rachel Carson's *Silent Spring* in 1962, drawing attention to the detrimental environmental effects of DDT and other pesticides.[4] Obviously, there are intermediate positions that view environmental problems and concerns as modern but that trace the beginning of modernity to some date earlier than 1960.

Which of these positions we adopt often makes a significant difference in the way our responsibility to the environment is conceptualized.[5] Those who would trace our environmental concerns back to Asoka or even earlier, and who envision a continuous history therefrom, are likely to view the environment as a resource. There is a trivial sense in which this view is correct: without a properly behaving environment, we would not exist as a species. We need a functional physical environment in order to survive. However, there is also a very obvious, but nevertheless nontrivial, sense in which it is correct. This sense is nontrivial because, when properly explicated, it shows

[2] Sukumar (1994) describes some of the relevant texts that are yet to be translated into any modern language.

[3] See, for example, Haila and Levins (1992), p. 1.

[4] Carson (1962); Lear (1997) provides a discussion of Carson's impact.

[5] Trivially, there is also a dependence in the other direction: how we regard our responsibility to the environment influences how we reconstruct the history of environmental concern. The two dependencies reinforce each other.

that regarding the environment as a resource does not demean it. An analogy will be useful here. We need more than a properly functioning physical environment in order to survive as a species. Without properly functioning reproductive partners, for example, we would not exist as a species.[6] Presumably, as individuals, we should not view reproductive partners *only* as resources. The emphasis here is on the *only*;[7] there is nothing wrong, even if it is psychologically disquieting, in viewing such partners in part as resources for our biological functioning. We use their bodies, as they use ours, for the function of reproduction. Presumably, unless we have completely divorced reproduction from the type of subjectivity associated with sex, which we may well do in an age of cloning and ectogenesis, these partners are also psychological and spiritual resources. Regarding reproductive partners in part as resources of this type does not demean them. If we are willing to regard reproductive partners *in part* as resources, environmental features can also be regarded as resources without demeaning them. These features may well be more than resources; but they are still justifiably regarded in part as resources. Regarding something as a resource in this sense does not imply disrespect.

From this point of view, concern for the environment consists of a worry that important resources will disappear if the environment further deteriorates. An analysis of what sort of value these resources provide will occupy much of Chapter 4; here it will only be assumed that there is some value that we do not want to lose. Biodiversity constitutes a resource in at least one obvious but uninteresting sense: unexpectedly useful substances are quite often discovered from living organisms. One of the best-known, though perhaps not representative, examples is the discovery of alkaloids in the Madagascar periwinkle (*Catharanthus rosea*), which resulted in better treatment of a suite of diseases from hypertension to Hodgkin's disease and leukemia. Because of these compounds, there is now a 99 percent probability of remission of lymphocytic leukemia if properly treated.[8] Had the Madagascar rainforest vanished before this plant was discovered, we would

[6] For the sake of this argument, reproductive partners are not being considered as part of the environment. In this argument, what is being regarded the environment of an individual is its environment *qua* species.

[7] Those already familiar with philosophy will recognize a Kantian theme here: not regarding another individual only as a means, but also as an end, does not imply that that individual cannot ever be regarded as a means.

[8] See Caufield (1984), p. 220, for a discussion of the discovery of the medicinal benefits of *Catharanthus rosea*; Lewis and Elvin-Lewis (1977) provide botanical details and other similar examples.

have lost this resource. In this sense, biodiversity is obviously a resource. But, as we shall see in Chapter 4, there are more nontrivial senses in which biodiversity is also a resource for us. What is most troubling about the loss of biodiversity is that potential resources are being lost before they are even discovered; many species are becoming extinct before they have even been identified, let alone studied. Section 2.1 will explore the nature of the worry about the loss of such resources, and relate it to the influence of a powerful myth that was earlier dubbed the myth of lost futures.

By contrast, claims that our environmental concerns are radically new are usually accompanied by an emphasis on those aspects of environmental features that are supposed to make them more than mere resources.[9] According to this position, treating these features as resources demeans their "intrinsic value"; sometimes this position forms part of the ideology of "deep ecology." Intrinsic values and deep ecology will be discussed in some detail in Chapter 3 (§§ 3.2–3.3 and § 3.4, respectively). This rather religious view of nature is often traced to a bewildering variety of sources, from Eastern mysticism to indigenous American practices (see § 2.2). Western man – and many of those who hold this view prefer such a gendered formulation for political reasons – is supposed to have lost something that these other societies still possess, and he has only discovered this loss, and its significance, very recently. This loss is variously attributed to technology, to social evolution from a primeval state to patriarchy, sometimes to capitalism itself, and to many other factors. What these etiological sources generally share is their association with the emergence of the modern world, a condition variously identified as modernity, enlightenment, scientific rationality, and so on. Now Western man must make up for lost time and prevent the further desecration of nature before there is none left. A deep sense of loss accompanies these beliefs. The relevant myth in this case was earlier called the myth of the golden age, the age from which Western man has allegedly fallen to his present state.

A touch of skepticism is in order here. Most of the positions mentioned in the last paragraph are typically presented with great rhetorical flourish and intense emotive appeal but with little consideration for their cognitive or empirical content. For instance, how can one not have the greatest reverence for a majestic thousand-year-old tree? We often have an unreflective preference for the natural over the artificial, until we begin to wonder whether we should have any less reverence for a magnificent thousand-year-old fort

[9] See Borgmann (1995), which will be further discussed in § 2.2.

or temple.[10] For those who want reasons beyond rhetoric to inform their actions, routine recourse to rhetorical devices is often unpalatable. Worse, as we shall see, sometimes deeply interesting phenomena are lost, or at least not noticed, because of rhetorical excess. There is indeed something very interesting, worrisome or not, about our relationship with technology. It is quite likely that the unprecedented rapid development of technology in the second half of the twentieth century marks a qualitative change in our relation to the world compared to all earlier generations. Quantity gets transformed into quality, as Engels once pointed out: the relentless improvement of technology during the last century may have made the nature of technology qualitatively different from anything humans have ever had access to before. One result of this change in technology has been the increasingly rapid transformation of the environment. Mechanized logging since 1960 has contributed to the reduction of Costa Rica's forest cover from 55 percent in 1961 to 22 percent in 1991.[11] Large dams have destroyed biologically interesting habitats almost everywhere they have been constructed.[12]

Twentieth-century technology has even radically transformed our subjective phenomenological experience. We may see loggers destroying rainforests on our television screens as they are doing it. We can see rainforests burning. On 9 September 1987, a U.S. National Oceanic and Atmospheric Administration satellite, NOAA-9, flashed back images of 7,603 fires then burning in the Amazon basin in Brazil, 2,500 of them in the single western state of Rondônia.[13] There is the potential for immediacy in our response: many individuals feel the urge to stop them, the loggers and the fires. There is no longer the luxury to think, reflectively, that what is being observed now actually happened a while ago, and that, while such events may be unfortunate, nevertheless all that we are left with is the option of carefully considering policy alternatives in order to prevent future destruction.

Strangely, many among us do not seem to have the same reactive response as we see cities being bombed as it happens. In the United States, television images of the calculated destruction of Baghdad and the loss of innocent life, now accomplished twice in our lifetimes, apparently do not affect most individuals in the way they are affected by the loss of a rainforest. In the case of war, technology seems to have the opposite effect: by mechanizing killing,

[10] Sober (1986) makes a similar argument.

[11] The statistics are from Faber (1993), p. 138. See also Sader and Joyce (1988) and Harrison (1991).

[12] McCully (1996) provides comprehensive documentation of the devastating environmental effects of dams around the world.

[13] Revkin (1990), pp. 231–233.

it lets us avoid the phenomenology of hand-to-hand combat, that other experience of killing. The variety of subjective responses to technology merits much more analysis than has so far been afforded to it. But any such attempt is well beyond the scope of this book.

2.1. THE MYTH OF LOST FUTURES

Even worries involve assumptions that can be uncovered by systematic excavation: by figuring out exactly why, in worrying about a specific entity, we are reacting to the world in the way that we do. The view that our concern for the environment is ultimately a concern for highly important resources forms part of at least two types of ideology that frame the relationship between our human existence and the world around us: (i) a scientific (or intellectual) ideology, and (ii) an economic ideology. The *first* ideology – and "ideology" is not intended to have a negative connotation – was explicitly articulated by the ecologist Daniel Janzen in his justly famous 1986 call to action, "The Future of Tropical Ecology." In this piece, which forms one of the founding documents of conservation biology, Janzen bemoaned the wanton destruction of neotropical rainforests during the previous generation and emphasized the urgency of conservation: "Within the next 10–30 years (depending on where you are), whatever tropical nature has not become embedded in the cultural consciousness of local and distant societies will be obliterated to make way for biological machines that produce physical goods for direct human consumption."[14] Janzen argued that "biologists are in charge of the future of tropical ecology."[15] He noted the direct utilitarian benefits that tropical ecosystems can provide for human societies: "How many potential polio victims realize that their vaccine was grown in a chicken egg, and chickens are nothing more than tropical pheasants specialized at preying on bamboo seed crops (which an Illinois farmer mimics with his chicken feed)?"[16] But he also noted the intellectual

[14] Janzen (1986), p. 306. When speaking of the need to embed concern for the habitat into local consciousnesses, Janzen was presumably referring mainly to Costa Rica, where he worked. Unlike the situation in most of the neotropics, indigenous traditions that place a high value on their forests have disappeared in Costa Rica. In a typical neotropical context there is no problem of culturally embedding such values, at least for indigenous peoples, if not for national elites descended from foreign invaders and occupiers. The problem of overconsumption remains valid in all cases.

[15] Ibid.

[16] Ibid., p. 316.

stimulation that the diverse tropical ecosystems provide, not only for ecologists coming from outside but also for people living in and around them. The piece included a famous exhortation:

> If biologists want a tropics in which to biologize [*sic*], they are going to have to buy it with care, energy, effort, strategy, tactics, time, and cash. I cannot overemphasize the urgency as well as the responsibility. . . . If our generation does not do it, it won't be there for the next. Feel uneasy? You had better. There are no bad guys in the next village. They is us [*sic*].[17]

The *second* (economic) ideology lies behind, for instance, the work of environmental economists who attempt to defend environmental protection on economic grounds.[18] This takes two major forms: (a) attempts at a calculation of costs incurred because of harm to the environment (including the loss of biodiversity), and (b) attempts at a calculation of benefits that would be lost (forgone opportunities) owing to a negative change in the environment (once again, including the loss of biodiversity).[19] From these perspectives, such calculations become important for us because environmental resources are limited. Prudence then requires us to curtail consumption in some cases. In most circumstances, this limitation also forces trade-offs between different goals. Calculations such as these are usually performed in the economic context of individual-based capitalism. However, as the political philosopher G. A. Cohen has pointed out, traditional Marxism also assumes that environmental resources are unlimited, and that technological innovation can continue to deliver ever-increasing supplies of material goods and services.[20] Ecological limitations dictate against such unbridled optimism, which Cohen takes to be one of the factual assumptions in Marxism that should now be rejected. According to him, Marxists must modify what they suggest as the proper politics of the future: the proletariat is no longer marching inevitably toward communism and into a future of continuous material progress.

What these ideologies – and many others – share is a fear of a future with many fewer available options. One way to understand the background

[17] Ibid., p. 306.

[18] See, for instance, Pearce (1993) for an early introduction.

[19] Conserving biodiversity may also involve forgone opportunities. This will be discussed in Chapters 5–7. It will be argued in Chapter 7 that these calculations are often little more than arbitrary. Whether or not these calculations should be taken as indicating anything useful is irrelevant to the point being made in the text.

[20] Cohen (2000), pp. 104–105.

assumptions common to all of these ideologies is to elaborate a very general picture or myth that is supposed to modulate conscious ideologies into taking the forms that they do. Unlike an ideology (as that term is being used here), a myth is not necessarily something that any individual consciously believes or could even fully articulate if asked; moreover, it may be the case that no individual accepts all features of the myth.[21] The term "myth" is not being used here in the customary sense of anthropology. Rather, it is being used in a sense that is more continuous with everyday language. As used here, a myth is a general story with normative[22] implications, parts of which are known to be false, or at least implausible, but which is nevertheless useful in analyzing other, more veridical stories that share some crucial aspects with it. If we were doing the epistemology of an empirical science, an idealized model would have a function similar to that of a myth in the present context.[23] Just as the same ideal model can be used to analyze a variety of situations more closely connected to experimental results, the same myth can be used to analyze many conscious ideologies. "Myth" is not being used pejoratively. Nevertheless, it is intentionally being used to remind us that some of its ingredients are nonveridical. This is important, because individuals operating under the aegis of a myth may act as if its literal truth is a given.

In this context, what unites the biologists and the ecological economists mentioned earlier is a myth of lost futures. Most scientists concerned with biodiversity worry about the loss of species, communities, and other biological entities that they study. For many scientists, what is perhaps even more psychologically troubling is the potential loss of the wonder one feels when confronted with a new form of life. In Australia in the 1970s, frogs were discovered that undergo their entire development inside the mother's stomach after a female swallows its fertilized eggs.[24] In the 1990s, a large

[21] If no convincing myth of this sort can be found, we have misclassified our ideologies in the first place. There may be nothing relevantly common to the set of ideologies we have categorized together.

[22] Here, "normative" is used in the axiological sense, being restricted to ethical, political, or aesthetic considerations; epistemological normativity is being excluded.

[23] For instance, a model in population ecology may assume that predators and prey encounter each other at a rate directly proportional to both of their densities, an assumption that is known to be an idealization. (See also Chapter 5.)

[24] The two species that did this are *Rheobatrachus silus* and *R. vitellinus*, the so-called gastric brooding frogs that were endemic to the rainforests of southern Queensland. Both species became extinct during the 1980s (McDonald 1990). They will be further discussed in Chapter 3 (§ 3.2).

new ox species[25] and three new species of barking deer[26] were discovered in Vietnam, while yet another species of barking deer was discovered in Burma.[27] The loss accompanying the disappearance of unknown species is perceived as a very tangible and painful loss by biologists.

Perhaps even more important, careful observation of novel ecological communities was not only critical for framing many ecological generalizations, it was even more critical for the formulation of the theory of evolution by natural selection. Darwin began to understand the process of evolutionary divergence through small blind variations by thinking about the etiology of the geographical patterns of variation in the beaks of the Galápagos finches.[28] Independently, Wallace obtained an even clearer conception of the principle of divergence of taxa from common ancestors while wandering around the islands of the Malay archipelago (in contemporary Malaysia and Indonesia).[29] He was collecting specimens for a living, but more importantly, he was pondering upon the geographical and successional contiguity of closely related species. The observed continuity of forms drove home to Wallace the idea that successive small variations, if preserved and amplified through selection, can explain divergences between species and other higher taxa. No one knows what new discoveries can be made in the unexplored habitats, in some tropical forests, for example, or in the oceans. Biologists such as Janzen feel that a book is being permanently destroyed just as they are beginning to read it. For many such field biologists, and for others with a penchant for exploring new habitats, their relationship to biodiversity is not just intellectual: it is a source of spiritual solace. Wonder and reverence marks their relation to biological complexity and diversity. Wonder and reverence is part of what a good life is for most of us, not just for biologists. What generates wonder and reverence is not a trivial resource. The future may have none of it.

For ecological economists, similarly, the prospect of a degraded future is often just as depressing. Many believe that in a tangible sense the future

[25] This is the saola, *Pseudoryx nghetinhensis* (Robichaud 1998).

[26] These are the giant muntjac, *Megamuntiacus vuquangensis* (Schaller and Vrba 1996); the Truong Son muntjac, *Muntiacus truongsonensis* (Giao et al. 1998); and the Pu Hoat muntjac, *Muntiacus puhoatensis* <http://ebooks.whsmithonline.co.uk/encyclopedia/42/ M0008442.htm>.

[27] This is the leaf muntjac, *Muntiacus putaoensis* (Amato, Rabinowitz, and Egan 1998). For a popular account of its discovery, see Rabinowitz (2001).

[28] The history of Darwin and his finches has been repeated many times; see Sulloway (1982) for an important attempt to separate legend from "reality."

[29] On Wallace's use of biogeography, see Sarkar (1998c) and Voss and Sarkar (2003).

is likely to be poorer than the present. For the economist David Pearce, following Kenneth Boulding, the contrast is between the "cowboy" economy of today, with unbridled consumption, and a "spaceship" economy in which the finitude of resources is appropriately taken into account.[30] Two conclusions are supposed to follow immediately from this assumption: (a) the current rate of consumption, which assumes infinite resources, cannot be sustained – we will have to learn to consume less; and (b) we have to make choices between possible alternatives in our consumption patterns. Overconsumption leads to a reduction of such choices as some resources are irretrievably lost. Conclusion (a) is the direct sense in which the future is likely to be poorer than the past. Conclusion (b) shows perhaps a more troubling way in which the future will be bleaker: the variety of our pleasures, intellectual and physical, will be more limited.

How seriously should we take these arguments? To what extent is the myth of lost futures based on a reliable extrapolation from our admittedly troubled present to the future? That a loss of unknown species will occur if unexplored biologically rich habitats such as tropical rainforests and coral reefs disappear hardly seems doubtful. However, how likely is it that this is going to lead to the disappearance of truly novel forms of life, at higher taxonomic levels than species and genera, that would sustain our intellectual curiosity? The truth is that we do not know. We may point to the continued discovery of novelty in the past. Skeptics will point out that our history of exploration of this sort is a very short one compared to all of human history. Biological science, as we practice it, goes back no further than Linnaeus (1707–1778). Perhaps we have already described almost all of the truly diverse living forms, for instance, most of the different phyla. Skeptics will urge that we *may* have reached a situation of diminishing returns. Not only does more effort result in decreasing rewards, we *may* have so few potential new rewards that continued effort is nonoptimal. As we become more familiar with the biodiversity around us, according to these skeptics, it will generate less reverence and wonder. Moreover, there may not be enough new discoveries in the offing to justify preventing other potentially valuable uses of land.

What about biological theories? Skeptics of the value of biodiversity conservation may argue that only one truly important theory has emerged from observation of the variety of nature, the theory of evolution by natural selection. This is perhaps the single most important discovery in the history of biology, but it is the only one in which the diversity of observable

[30] Pearce (1993), elaborating on Boulding (1966).

life was crucial. From such skeptics' point of view, what little we know of theory in ecology (and we do not know very much – see Chapter 5) did not require such diversity. Observing a small typical set of systems would have been sufficient: there are only a few distinct types of ecological dynamics.[31] Moreover, the future of biology seems increasingly to lie in the laboratory and the computer and not in wandering around forests. At least sociologically, natural history is no longer at the center of biological science. A future of continued spectacular advances in biology based on detailed observation of the bewildering variety of ecological systems in the field may well be purely a myth with no veridical ingredient. Skeptics will urge that our resources not be directed toward the conservation of such systems at the expense of other human needs and desires.

Nevertheless, a prudential argument can be made in favor of biodiversity conservation. The extinction of species, communities, and other components of biodiversity is irreversible.[32] Consequently, prudence dictates that, in the presence of uncertainty, we should conserve all such components of biodiversity.[33] This is a powerful argument, but still not a definitive one. Biodiversity conservation may have to come at the cost of other alternatives which may also present irreversible choices. Even a forgone economic opportunity may be irreversible. For instance, in an underprivileged forest community, the loss of immediate revenue from logging and other extractive activities may do irreparable harm to the future of that community. If the community is truly rendered incapable of economic survival, this harm is not only long-term, but also irreversible. Meanwhile, if the associated potential economic gains are high, prudence no longer clearly dictates the conservation of biodiversity. All we may be left with finally is a worry that we will lose the opportunity for the kind of spiritual solace that biodiversity provides for those who have daily contact with the variety of the natural

[31] If ecology followed this route, one could argue, it would join the rest of biology, which, since the 1980s, has increasingly focused on a small group of "model" organisms (or systems) that are supposed to explicate biological universals by their representativeness. For an introduction to controversies over the use of model organisms in evolutionary biology, see Kellog and Shaffer (1993); for developmental biology, see Bolker (1995). These sources provide an introduction to a large and growing literature.

[32] Though this degree of precision is probably unnecessary, throughout this book "component" of biodiversity will be used to indicate some structural unit (in either the ecological or the taxonomic hierarchy) that displays variability and, therefore, contributes to biological diversity; "feature" of biodiversity will be used to refer to some property that is valuable for the same reason, for instance, rarity or endemism; "aspect" will be neutral with respect to these two categories.

[33] This is one version of what is sometimes called an argument from ignorance. Recall its connection to the rivet argument (Chapter 1, § 1.3).

world. This is less than satisfying. Moreover, nothing precludes the skeptic from arguing that even the physical experience of contact with nature can be technologically simulated.

The argument from ecological economics assumes that technological possibilities are not unlimited. It is, so to say, the pessimists' argument. Optimists will express an inexorable faith in human ingenuity.[34] Neither side doubts that resources are limited in the physical sense: there is only so much of each kind of material or process on Earth. But what should properly count as resources are materials and processes as they are utilized by us (and, depending on the context, by other species) for specifiable ends. This depends on the available technology. Technological innovation will convert what are today's nonresources into tomorrow's resources. Thus, according to the skeptics, not only is technological progress effectively unlimited, it also makes resources unlimited. There are many reasons to doubt such an overly optimistic prognosis. For instance, after several generations, we still have no technological solution to the problem of nuclear waste. Nevertheless, there have been many unexpected technological innovations, and we cannot answer the skeptic with complete certainty. Economic optimists and pessimists are both trading in uncertainties.

Whatever we do, we are forced to proceed under a shroud of ignorance. This calls for caution. Coping with uncertainty, probably intrinsically unquantifiable uncertainty, is one of the problems broached in Chapter 7 (§§ 7.4–7.6). It will not be solved successfully; the inevitable uncertainty of the future will continue to provide another subsidiary argument against unbridled technological optimism.

2.2. THE MYTH OF THE GOLDEN AGE

Far more troubling than the myth of the lost futures is the myth of the golden age. Characterized broadly, and glossing over many historical caveats, this myth goes back at least to Rousseau and the eighteenth century. Human society is supposed to have made a transition from a savage state to an initial communal state that brings out the best in all humans. All modern societies have fallen from that state. For Rousseau, private property in the form of enclosed land epitomized the transition from the communal to the fallen state. This vision of social change formed part of Romanticism, the Romantic ideal being the golden age of the first civil societies. It is a

[34] For an extreme version of optimism, see Easterbrook (1995).

vision of a fall from grace. "Civilization," from this point of view, has meant a continuous descent. Reason, the source of science, technology, and the other amenities of civilization, is one of the forces allegedly responsible for this degradation of humanity from its primeval state.

In the environmental context, this means that economic development and technology, which are allegedly connected to instrumental (or resource-exploitative) reason, are supposed to be among the sources of evil. Those who act under the aegis of the myth of the golden age often – but not always – live in the affluent North, enjoying the results of the highest degree of technological development seen in history.[35] Nevertheless, the myth of the golden age makes them question the value of such lives. Some of them are often convinced that their own cultural resources are deficient and will not allow them to recover what has been lost. They hearken for alleged wisdom from other sources. For some, what has been lost is the balance of nature that once existed between different natural features and humans.

Here is a typical sentiment supposedly expressed by a Buddhist monk from Thailand:

> The times are dark and *siladhamma* is asleep, so it is now the duty of monks to reawaken and bring back *siladhamma*. Only in this way can society be saved. *Siladhamma* does not simply mean 'morality' as commonly supposed. You may obey all the *silas* (commandments) in the book, it still doesn't mean that you have *siladhamma*, which in truth means 'harmony,' the correct balance of nature, the natural result of natural harmony.[36]

To take another example: Albert Borgmann claims that in "the original human condition[,] . . . as far as we know, both the nature of reality and the reality of nature were divine. The human attitude that corresponded to this unified world was one of piety. . . . Here on this continent [America] we are at least vicariously within hailing distance of our original condition thanks to the heritage of the Native Americans."[37] Non-Western religions and cultural practices retain insight into the golden age because they reject the type of civilization that is associated with the technological progress of the West. Indigenous groups continue to have a grip on the golden age

[35] Deep ecologists (see Chapter 3, § 3.4) are among those who exemplify this position.

[36] Batchelor and Brown (1994), p. 90. Note that the objectives allegedly expressed by this monk remain anthropocentric.

[37] Borgmann (1995), pp. 31–32. As is typical in such discourse, all native American ethnicities (the "First Nations" of the Americas) are lumped into a single category, and *all* are supposed to be such repositories of environmental wisdom.

because of their lack of technological development. Modernity is rejected in both cases.

The myth of the golden age is relevant to discussions of biodiversity conservation in two conflicting ways: (i) it potentially provides a rationale for the preservation or re-creation of wilderness in its pristine state (without any permanent presence of humans) – this will be discussed in detail in the next section; and (ii) it potentially argues for the value of including indigenous (and, sometimes, other local) residents of an area targeted for biodiversity conservation into management plans, and for allowing them to continue their traditional practices. The assumption that leads to this conclusion is that these inhabitants live in harmony with nature, like Borgmann's Native Americans, and their activities are not detrimental to the persistence of biodiversity. The second position is of considerable value because it provides a much-needed antidote to the conflation of wilderness preservation with biodiversity conservation that has sometimes been inimical to the interests of biodiversity (see the next section). Moreover, it incorporates a welcome respect for other cultural practices, respect often lacking in the affluent North. Nevertheless, the question of how a particular group with a given set of practices will affect the environment is an empirical question (and a well-studied one), as was pointed out toward the end of Chapter 1 (§ 1.3) during the discussion of the tragedy of the commons argument. An assumption that all indigenous and other groups with premodern habits have ecologically sound practices is yet another Romantic myth (and one that is almost always fictitious) – that of the "ecologically noble savage."[38]

What makes the myth of the golden age particularly troubling is not just the conflation of wilderness preservation with biodiversity conservation, or a belief in the myth of the ecologically noble savage – even though both assumptions, when generalized to all situations without empirical confirmation, may have adverse consequences for biodiversity (as the next section, in particular, will document). This myth is particularly troubling because of five assumptions that usually accompany it (even though they are not logically connected to it in the sense that an acceptance of the assumptions embedded in this myth logically entails the acceptance of these other claims):[39]

(i) that those groups that occupy some habitat over a long period of time have a deep knowledge of that habitat. Going further, this point is

[38] See Redford (1990) for further discussion of this myth.

[39] In fact, whatever entailment there is holds in the other direction, as will be shown later in the text.

sometimes expressed by saying that they have an "organic relation" to that habitat;
(ii) that those groups that have the relation described in (i) with their habitats are ethically superior and, therefore, deserve more approbation and greater respect than those that do not;
(iii) that technological advance, guided by science and reason, is inherently deleterious to the quality of the environment;
(iv) that technology is morally deplorable – going further, for the sake of argument, that science is morally deplorable; even further, that reason and rationality are morally deplorable; and
(v) that groups advocating science or rationality as part of their cultural ideology are inferior to those that celebrate the primacy of raw nature.

Assumptions (i) and (iv) entail most of the ingredients of the myth of the golden age. It is often implicitly assumed (for instance, in Nazi doctrine – see below) that (i) entails (ii), that (iii) entails (iv), and that (iii) or (iv) alone (and, of course, both together) entail (v). However, except perhaps for an entailment of (v) from (iv) (and, therefore, also of [v] from both [iii] and [iv]),[40] there is no such entailment (if we are genuinely concerned with logic): (i) and (iii) are strictly descriptive claims, to be decided by empirical data, no matter how difficult that determination may be in practice; (ii), (iv), and (v) are normative claims that cannot follow from strictly descriptive claims.

How should we treat these assumptions? Assumption (i) appears reasonable in its weaker forms as a statistical generalization from empirical data. Those groups or individuals with a long experience of some habitat often have an intimate knowledge of that habitat. Because of the complexity of most ecological systems, even seasoned ecologists often have much to learn from local experts when devising biodiversity conservation plans for a region. Consider just one example: in 1976, the evolutionary biologist Ernst Mayr recalled: "40 years ago . . . I lived all alone with a primitive tribe of Papuans in the mountains of New Guinea. These superb woodsmen had 136 names for 137 species of birds I distinguished (confusing only two nondescript species of warblers)."[41] The best scientific training in taxonomy – and Mayr had been trained in the world's foremost ornithological laboratory, that of Erwin Stresseman at the Museum für Naturkunde in Berlin – could add little to local knowledge. Nevertheless, if positing an "organic relation"

[40] Even this requires an additional, and perhaps controversial, assumption: that groups advocating morally deplorable positions are themselves morally deplorable.

[41] Mayr (1976), p. 517.

assumes some necessary connection between long experience of an environment and relevant intimate knowledge about its conservation, there is no such relation. Instead, there is an empirical question to be investigated: how often, and exactly under what historical and present conditions, does such a relation hold? This echoes a point raised earlier in this section and also in Chapter 1 (§ 1.3), that the status of the relation between human activity and the environment is empirical and must be empirically investigated rather than decided a priori. Devising conservation policy on untested intuitions about this relation is not only likely to lead to unsuccessful results but also, if it does harm to humans, ethically objectionable.

When we get to assumption (ii), we find bedfellows we may not appreciate. Assumption (i), the alleged entailment of (ii) from (i), and therefore (ii) were part of the Nazi critique of Jews, Romanos, and other groups with migratory histories.[42] These groups could not have an organic relation to the environment because of their migratory nature and were, therefore, inferior to the settled Aryan peasant. Meanwhile, Romantic ideology, including a dedication to the welfare of animals and the environment, was an important component of Nazi doctrine. "*Im neuen Reich darf es keine Tierquälerei mehr geben*," Hitler proclaimed with pride.[43] On 24 November 1933, the Nazis passed a strict law for the protection of animals (*Das Tierschutzgesetz*); this was followed on 3 July 1934 by a law limiting hunting (*Das Reichsjagdsgesetz*), and finally, on 1 July 1935, by the most comprehensive nature protection legislation that had ever been enacted anywhere in the world (*Das Reichsnaturschutzgesetz*). It is the organic relationship with the land (assumption [i]) that led to such concern for the welfare of the environment. According to Nazi doctrine, the Jews also fell afoul of assumption (v). Nazi doctrine is not very clear about the status of assumptions (iii) and (iv), and with good reason: a worship of technology was also part of Nazi ideology, living in very uneasy tension with the Romanticism. However, high culture, as indicated by certain types of music and science, and exhibited most prominently by Germany's Jews, represented moral degeneration to the highest degree.

We have to be very careful here. First, sending human beings to prisons or concentration camps, let alone to death camps, is *not* entailed by assumptions (i)–(v). It is important not to impute a potential for such crimes,

[42] For details of the positions mentioned in this paragraph, see Biehl and Staudenmaier (1995). See also Ferry (1995).

[43] "In the new Reich cruelty towards animals should not exist"; quoted from Ferry (1995), p. 91.

even by suggestion, to those environmentalists who are guilty of nothing more than being enamored of Romanticism. Second, the simple fact that the Nazis and some other groups share some assumptions does not automatically discredit those other groups and their programs. What we must try to understand is exactly what it is about these assumptions that make them congenial to all such doctrines, that of the Nazis and those endorsed by these other groups. Principally, it is the rejection of what is often labeled "humanism" and what that entails.[44] Humanism puts human interests at the center of what is morally and intellectually desirable. A very extreme form of a rejection of humanism would see human culture, generally, as a source of evil. From such a position, it is but a short step to have a general dislike of all humans who are different from those satisfying the demands of the Romantic ideal. The humanism of the Enlightenment brought with it a conception of individuality that held liberty as sacred and reason as the source of all that is valuable in culture. Obviously, the Nazis found these values troublesome in the pursuit of their political agenda. Many environmentalists who invoke the myth of the golden age also find human interests and human cultural (including scientific and technological) progress distasteful. Beyond this – perhaps accidental – commonality, no stronger conclusion should be drawn from the facts noted earlier about the values held by these environmentalists. Nevertheless, in Western environmentalism there continues to be a radically right-wing strand with a strong dislike of humanism. One particular form that this takes, "deep ecology" and the associated ideology of biocentrism, will be briefly discussed in Chapter 3 (§ 3.4).

2.3. WILDERNESS

The golden age is also associated with another pervasive myth of great emotive force among the affluent: the myth of untouched wilderness, natural landscapes in what are taken to be their primeval states. "Nature" in the United States, according to Borgmann, "speaks . . . in the ancient voice of the wilderness";[45] thus the existence of wildernesses is explicitly linked to

[44] On "humanism" and its commitment to equality (before the law) of all individuals, irrespective of accidents of birth (ethnicity, gender, etc.), see Ferry (1995). Ferry emphasizes humanism's commitment to rights and responsibilities for all human individuals, but *only* for human individuals.

[45] Borgmann (1995), p. 42.

the golden age. Feminists have very good reasons to deconstruct many myths associated with wilderness, such as that of "virgin" forests, often found in naturalists' accounts, which suggest a peculiarly masculine relation of the observer to what is being observed, of those who enjoy to that which is being enjoyed; such a deconstruction, though undoubtedly of great political value, is beyond the scope of this book.[46]

Analyzing the wilderness myth is particularly important in the context of this book because wilderness preservation is routinely conflated with biodiversity conservation, especially in Northern societies. This conflation is so pervasive that even the Convention on Biological Diversity adopted at the Rio de Janeiro United Nations Conference on Environment and Development (the 1992 "Earth Summit") explicitly mentions wilderness as a conservation target for biodiversity. Annex I of the convention contains a list of targets for identification and monitoring. The first entry lists as some of the targets of conservation:

> Ecosystems and habitats: containing high diversity, large numbers of endemic or threatened species, or wilderness; required by migratory species; of social, economic, cultural or scientific importance; or which are representative, unique or associated with key evolutionary or biological processes.[47]

This formulation suggests that wilderness is a biological category because it is invoked – along with high diversity and the presence of endemic and threatened species – in the first, biological clause of the formulation, and not the third, mainly cultural clause. For this classification to be adequate, there should be some relevant *biological* similarity between the Alaskan tundra (a wilderness) and a neotropical rainforest (with high diversity and endemism of species).[48] In the United States, the Wildlands Project is committed to biodiversity conservation through the reconstruction of wildernesses.[49]

In sharp contrast, customary colloquial uses of "wilderness" refer to the absence of humans rather than to any special biological property. Indeed, it is this use that is incorporated into one of the most important pieces of

[46] For the beginnings of such an analysis, see Plumwood (1998).

[47] <http://www.biodiv.org/doc/legal/cbd-en.pdf>.

[48] Moreover, neotropical rainforests are generally quite heavily populated and do not satisfy the definition of "wilderness" that will be given in the following text. This example is being used here because advocates of wilderness preservation often act as if rainforests, and all the other pieces of biodiversity they want to preserve, are wildernesses. This move illegitimately ignores the often definitive role played by humans in creating these habitats (see the following discussion). See Sarkar (1999) for more detail on this point.

[49] For details on the Wildlands Project, see Soulé and Terborgh (1999).

legislation adopted anywhere for wilderness preservation, the 1964 U.S. Wilderness Act:

> A wilderness, in contrast with those areas where man [*sic*] and his own works dominate the landscape, is hereby recognized as an area where the earth and its community of life are untrammeled by man, where man himself is a visitor who does not remain. An area of wilderness is further defined to mean . . . an area of undeveloped . . . land retaining its primeval character and influence, without permanent improvements or human habitation, which is protected or managed so as to preserve its natural conditions and which (1) generally appears to have been affected primarily by the forces of nature, with the imprint of man's work substantially unnoticeable; (2) has outstanding opportunities for solitude or a primitive and unconfined type of recreation; (3) has at least five thousand acres of land or is of sufficient size to make practicable its preservation and use in an unimpaired condition; and (4) may also contain ecological, geological, or other features of scientific, scenic, or historical value.[50]

Clause (4), at the end, underscores the point that scientific features, though perhaps desirable, are not part of the necessary conditions defining a wilderness. Even when there may be a scientific reason for designating a place as wilderness (under clause [4]), there is no explicit relation to biodiversity. Moreover, clause (2) notes the important point that wildernesses deserve protection because they serve a deep spiritual need for many, at least among the overconsumptive affluent. Some of the affluent may just need solitude, which is less and less available as more and more land succumbs to the expansion of human industry and settlement. Others may find inspiration in seeing nature that cannot be tamed by human technology; the emotions that such an experience invokes places it in the philosophical category of the sublime,[51] which, like the beautiful, should command our respect. These are important human aesthetic values.

"Wilderness," as a category of positive concern – as opposed to "waste" lands to be tamed and used efficiently by humans – is of recent and highly localized vintage. As the historian Roderick Nash put it: "Friends of wilderness should remember that in terms of the entire history of man's [*sic*] relationships with nature, they are riding the crest of a very, very recent wave."[52] The origins of this concept of wilderness go back again to Romanticism; pristine wildernesses formed part of what has been lost from the golden age as human values are supposed to have replaced "natural" ones. However, the

[50] <http://www.fs.fed.us/outernet/htnf/wildact.htm>, § 1131 (c).

[51] On this use of "sublime," see Cronon (1996b).

[52] Nash (1973), p. xii.

use of "wilderness" that is most relevant in the present context of concern for biodiversity emerged only during the late nineteenth and early twentieth centuries and, according to most historians, initially in the United States.[53]

In the United States, wilderness preservation may be taken to have started with the passage of legislation to create national parks, Yosemite being so declared in 1864, with Yellowstone following in 1872. In retrospect, what is most ethically troublesome about the designation of such areas as uninhabited wildernesses is that they were generally created by the deliberate and forced expulsion of human residents (in this case, the First Nations) and an erasure of their histories. Here "erasure" refers to the systematic, if unconscious, reconstruction of societal memory in order to recast as uninhabited "wildernesses" the lands from which the original inhabitants were forcibly expelled. The Yosemite nation was excluded from Yosemite, the Crow, the Shoshone, and the Bannock from Yellowstone.[54] The final stage of exclusion was achieved at the end of the last "Indian" wars, when the remnants of the First Nations were herded into reservations and their traditional lands were declared to have been unoccupied by humans from the beginning of time.[55] These so-called wildernesses were hardly untouched by humans;[56] the same is true of the neotropical rainforests[57] and many other areas currently of interest for their high biodiversity content. It is this lack of veridicality that relegates the concept of wilderness to the category of yet another *myth* born of the golden age.

Biological criteria played almost no role in the designation of national parks in the United States until at least the 1940s. Rather, the national parks were sublime landscapes: mountains, waterfalls, and other landforms of exquisite and deep aesthetic appeal to transient visitors, who usually came from an urban elite rather than from the surrounding rural population. National parks began to be created in Canada in the 1930s and began to replace wildlife reserves in Africa and Asia starting in the 1940s.[58] Concern for general biodiversity, and not just for wildlife, began to play a role in

[53] For the connection to Romanticism, see Oelschlaeger (1991), Denevan (1992), and Cronon (1996a, b).

[54] See Spence (1999) for a detailed and sympathetic reconstruction.

[55] Cronon (1996b). Woods (2001) refers to these past practices as providing the "ethical" argument against wilderness preservationism and argues, correctly, that the ethics of preserving a particular potential wilderness area must be argued case by case.

[56] Denevan (1992), Cronon (1996b). Pyne (1982) documents the role of fire, often of anthropogenic origin, in the creation of these landscapes.

[57] Hecht and Cockburn (1990) discuss the Amazon basis; Sarkar (1999) provides examples from Central America.

[58] For more detail, see Sarkar (1999) and the references therein.

Table 2.3.1. ***Wilderness preservation and biodiversity conservation***

	Wilderness Preservation	Biodiversity Conservation
Objective	Landscapes without humans	Biological diversity at all levels of organization
Justifications	Aesthetic interest	Intellectual or aesthetic interest, present and future utility
Targets	National parks, wilderness preserves	High-biodiversity regions including national parks, preserves, community conservation areas, etc.; representative samples of biodiversity
Obstacles	Economic interests, overconsumption, human encroachment, invasive technologies	Economic interests, overconsumption, human encroachment, invasive technologies, habitat fragmentation, human exclusion (in some cases)
Strategies	Legislation, habitat purchase	Diverse methods including, but not limited to, legislation, habitat purchase

Note: For more detailed discussion of these issues, see the text and Sarkar (1999). This table is based on that source.

the design of national parks and other reserves only in the 1970s, after a recognition that all forms of life, and not just charismatic and useful species, deserve protection (see Chapter 6). It is at this stage that the legacy of wilderness preservation begins to play, at the very least, a questionable role when strategies for conserving biodiversity are being devised.

Table 2.3.1 compares wilderness preservation and biodiversity conservation with respect to (i) their objectives, justifications, and targets, all of which differ. However, the justifications for biodiversity conservation include the one that is typically put forward for wilderness preservation, namely, aesthetic value. The table also compares (ii) the obstacles and strategies they face, on which there is much common ground. These confluences – for instance, in having overconsumption as an obstacle – help to explain why these two projects tend to be conflated so often. However, once again, biodiversity conservation is more inclusive, embracing a wider spectrum of possible strategies than wilderness preservation. As Chapters 6 and 7 of this book will emphasize, designating national parks is not the only available strategy for biodiversity conservation. In many contexts, it may not even be the most desirable one. One such context is that in which there is local political opposition to human exclusion from a region that has traditionally been used for the resources it provides (see below). All that biodiversity conservation requires is the design and implementation of land management plans that do not harm biodiversity: this may well include some human use of the land.

The crucial point is that, for the purpose of biodiversity conservation, the effects of human presence and activity require empirical investigation. No general ban on humans is justifiable on empirical (or other scientific) grounds. Two examples (from many)[59] will drive this point home:

(i) In Keoladeo National Park in Rajasthan, India, a 450 ha. artificial wetland with shallow bodies of water was created by local rulers during the nineteenth century. This wetland attracts tens of thousands of wintering waterfowl and also supports large numbers of bird species that breed during monsoons.[60] Before Indian independence in 1947, the area was a hunting reserve (which is the purpose for which it was created) that also served as a grazing ground for cattle (mainly buffaloes) from the surrounding villages and as a water source for irrigation during the dry postmonsoon period. After independence, it was initially set aside as a bird reserve and, after 1981, as a national park.[61] On the advice of Indian and U.S. experts, who had not deigned to carry out systematic field studies, grazing was banned beginning in the early 1980s in an effort to promote bird diversity. When villagers protested the loss of fodder, the Indian state responded with violence and the police killed nine protesters. The ban on grazing, despite its intent, devastated Keoladeo as a bird habitat, especially for wintering geese, ducks, and teal. Paspalum grass and other opportunistic weeds, which had been kept in check by grazing, established a stranglehold on the wetland, choking the shallow bodies of water. Fish populations declined, leading to corresponding declines in bird nesting and populations. These results were presented in 1987 in a report of a ten-year ecological study conducted by the Bombay Natural History Society (BNHS), India's leading nongovernmental conservationist organization, which had originally promoted the ban.[62] The BNHS study was originally designed to substantiate the harm done by grazing. To its credit, the BNHS reversed its position in the face of recalcitrant data and began to promote the reintroduction

[59] For other examples, see, for instance, Stevens (1997) and Sarkar (1999).

[60] Gadgil and Guha (1995), pp. 92–93, provide an extensive discussion of the social background.

[61] See Lewis (2003), pp. 283–312, for details of the history.

[62] See Vijayan (1987), an unpublished report that documents these studies.

of grazing.[63] Officially, the policy has never been reversed, but it is no longer systematically enforced. As a result some (technically illegal) grazing continues, and anecdotal reports suggest that bird habitat has improved substantially, though the habitat would benefit from much more systematic grazing.

(ii) The second example comes from the Sonoran desert in Mexico and the United States. Two oases, on different sides of the U.S.–Mexican border, were subject to different management regimes. On the U.S. side, the protection of an oasis by its inclusion in the Organ Pipe Cactus Natural Monument led to a significant decline in species diversity over a twenty-five-year period. On the Mexican side, continued traditional use by Papago farmers at the Quitovac oasis, 54 km. to the southeast, produced no such decline.[64] There has been some controversy as to whether this difference in richness reflects biodiversity as measured, for instance, by indices such as those in Box 5.1.2 (p. 117). For birds, the situation is not clear. However, for plants, richness and diversity are both clearly greater at Quitovac.[65] Nevertheless, it has been argued that this diversity is not a sign of biological health; rather, the disturbance of the ecological community by farming allows many plant species to invade and also prevents the elimination of less-fit species by competitive exclusion.[66] Why this should count as a counterexample to the claim that human activity may help to maintain diversity is less than clear: as will be pointed out in the Chapter 5 (§ 5.1), it is almost a truism that competitive exclusion decreases diversity and simplifies ecological communities. Perhaps the intended argument is that even though human activity is resulting in higher levels of biodiversity, it is not the type of diversity that is desirable because it consists largely of "invasive" species. But even if this were true in Quitovac – and the evidence is not compelling – there is no ecological reason to believe that it is true in general. With respect to the biological composition of a community, it does not matter what the source of a disturbance is. What matters is its extent (or strength). There is no ecological reason to believe that some human disturbance can lead only to the establishment of "invasive" species but not the persistence

[63] Lewis (2003). Lewis also discusses a similar case in India, the presence of cattle in the Gir Forest (in Gujarat in western India), the last home of the Asian lion (*Panthera leo persica*). In this case, the cattle provide a substantial part of the food for the lions while not competing for the same vegetation species as nondomesticated ungulates in the reserve.

[64] Nabhan et al. (1982); see also Callicott (1991b).

[65] For the relevant definitions of diversity, richness, etc., see Chapter 5, § 5.1.

[66] An argument of this sort was made by Rolston (1991). For a reply, see Callicott (1991a).

> of native species that would otherwise have become extinct through competitive exclusion.

If biodiversity conservation is our goal, there is no avoiding the need for empirical studies, however difficult, to establish the cogency of management plans for land. The first example also shows how wilderness preservation through forced human exclusion, while depleting biodiversity (though only indirectly), may also come at a great human cost, providing the so-called ethical argument against wilderness preservation.[67] As should be expected, the human cost often generates political opposition, which can be inimical to biodiversity conservation. India provides many examples of such conflicts: for instance, in the Jharkhand region of east central India alone, thousands of villagers have been displaced since 1947 for the creation of national parks and reserves.[68] One result is that national parks and other reserves are not popular in that area, and poachers find it easy to recruit local residents. Forced displacement of people is ethically unacceptable for obvious reasons. However, this example underscores an additional prudential reason for avoiding such displacements as a means to create mythical wildernesses. If biodiversity conservation is not conflated with wilderness preservation, many such unnecessary political conflicts can be avoided. Above all, local support for conservation, which is desirable if not downright essential for success, may be retained. This is an important point and merits continued reemphasis: biodiversity conservation is as much, if not more, a political project as it is a scientific one.

[67] See Woods (2001) for this designation.
[68] Guha (1989), Agarwal (1992).

3

Intrinsic Values and Biocentrism

Why conserve biodiversity? This normative question has traditionally been part of the core of environmental ethics. In most of the discussions in the earlier chapters of this book – though not in Chapter 2, § 2.1 – we have simply assumed that we should attempt to conserve biodiversity. We attribute *value* to biodiversity, preferring its presence to its absence. But what is the source of this value? Why should we conserve biodiversity? We must be able to answer these questions if we are to have any confidence that we do not attribute value to biodiversity merely out of tradition, prejudice, or (what is much more plausible) fashion: biodiversity conservation *is* at present very fashionable in rich Northern societies. Moreover, answering these questions explicitly and convincingly is important above and beyond our having rational justification for attributing value to biodiversity. It has practical consequences, because arguments for biodiversity conservation must be made and evaluated in the political arena where decisions about the future of our biological heritage are made.

In that political arena, not everyone attributes value to biodiversity, or at least not to the same extent that biodiversity conservationists do. Among those who do not attribute value to biodiversity to the same extent that conservationists do are the owners and shareholders of the Japanese timber companies that have devastated Southeast Asian rainforests and are now moving to the neotropics;[1] the ranchers who are trying to convert Amazonia into pasture;[2] the loggers in the U.S. Pacific

[1] See Bevis (1995) for a semipopular account of the logging of Borneo and, with the disappearance of the great forests there, the attempts of logging companies to shift their operations to South America.

[2] See Caufield (1984) for a scintillating account; Marchak (1995) provides a global assessment of logging.

Northwest;[3] and, perhaps most importantly, the owners and shareholders of the multinational oil companies that are now prospecting in many places around the world, including the Alaskan tundra and neotropical rainforests (for instance, in the western Amazon).[4] If we are going to try to change the minds of these individuals, we must be able to provide good reasons. There *are* very good reasons, as we shall see (though more in the next chapter than in this). Nevertheless, there is also ample reason to doubt that good reasons alone will be successful in most cases, especially in arguing with those who worship immediate profit at the expense of everything else.

When argument alone cannot change the minds of those who would destroy biodiversity, we may have to turn to political strategies, including local actions such as protests and strikes, as well as to global actions such as consumer boycotts as part of our attempts to ensure biodiversity conservation. Since these strategies are necessarily at least in part coercive, we should have strong ethical justifications for our actions. Finally, and perhaps more importantly, we will routinely have to try to influence political institutions – most notably, governments, regulatory agencies, and international financial institutions – that have the power to control those who encourage or condone the wanton destruction of biodiversity by their actions (or inaction). In addition to exerting political pressure, we can best do this by providing convincing arguments and incontrovertible evidence to those who make the relevant decisions. Perhaps more efficaciously, we can enter into public debates and raise our questions and promote our concern for biodiversity. Once again, we would be wise to have good reasons and evidence.

This chapter will begin the exploration of possible reasons for conserving biodiversity; the following chapter will continue that exploration. These chapters are intended only as an introduction to some of the central issues in environmental ethics and not as a comprehensive treatment. Moreover, the treatment given here consists primarily of laying out the best arguments for a particular anthropocentric position and, in that sense, is not philosophically "neutral" (whatever that may mean). As noted earlier (in Chapter 1, § 1.2), an attempt has been made to minimize philosophical jargon in the interest of making the discussion accessible to a wide audience with no previous exposure to philosophical ethics. It will be assumed throughout that we

[3] See Norse (1990). See <http://pnwin.nbii.gov/nwfp/FEMAT> for a comprehensive study of the effects of logging on the Pacific Northwest.

[4] See <http://www.ran.org/oilreport/amazon.html>.

have an adequate enough understanding of what it means to attribute value to an entity and that we understand why attributing value takes us into the normative realm by suggesting certain courses of action over others. As the beginning of Chapter 4 will note in a little more detail, meta-ethical concerns, such as concerns about the sources of values in general, about whether only human individuals are capable of attributing values, and so on, are ignored here. An attempt has been made to keep the arguments of this chapter and the next neutral with respect to meta-ethical disputes. To the extent that meta-ethical and normative ethical claims can be sharply distinguished, this book will ignore meta-ethical issues, primarily because they influence conservation practice, if they do so at all, only through their influence on normative ethical claims.

In this chapter, the conditions that any set of reasons for conserving biodiversity – that is, a "conservationist ethic" – must satisfy in order to be judged as adequate will first be laid out explicitly (§ 3.1). There will then be an analysis of attempts to provide a justification for conserving biodiversity because it has value in and of itself, irrespective of whether it has any sort of value for humans. This strategy of justifying biodiversity conservation is based on attributing "intrinsic value" to nonhuman biological entities (§ 3.2). Though it has wide support among conservation biologists and some environmental ethicists, it will be found unsatisfactory (§ 3.3). Attributing intrinsic value to other species (and, often, to other biological entities) forms part of the related ideologies of biocentrism and deep ecology. These will also merit some discussion here (§ 3.4). Finally, there will be a very brief discussion of the unfortunate – and in many ways unexpected – tension between the ethics of the animal welfare and liberation and the ethics of biodiversity conservation (§ 3.5).

3.1. ADEQUACY CONDITIONS FOR A CONSERVATIONIST ETHIC

It is important, at the beginning of any discussion of an ethic of biodiversity conservation, to lay down the conditions that we require any such ethic to satisfy in order to be *adequate* before we begin to examine the status of any particular proposal. These conditions will spell out what *any* conservationist ethic must achieve. If we disagree about what these adequacy conditions are, we are more likely to disagree about the status of a proposed conservationist ethic than if we do not. In later sections of this chapter, and in the next, these adequacy conditions will be used to rule out many popular proposals for a conservationist ethic. These discussions will assume that we have agreement about these adequacy conditions.

Setting explicit adequacy conditions has an added benefit. Unless we have an explicit agreement about these adequacy conditions, the following unfortunate scenario – what amounts to a shifting of the goal posts – may occur. We start an argument for – or against – a particular proposal for a conservationist ethic assuming that we know the burden of proof. Suppose, for instance, we assume that our proposed conservationist ethic must encourage the protection of all species. However, provided that it is compatible with the goal of protecting these species, it may allow the desecration of pristine wilderness. Moreover, if wilderness preservation is incompatible with the protection of these species, it may even require the elimination of wilderness, for instance, through its conversion into human-designed habitat suitable for some endangered species. Suppose that our opponent is a closet "deep ecologist" (see § 3.4) committed more strongly to the preservation of wilderness than to the protection of all species. At a stage when, from our point of view, the argument should be over and we should begin consideration of strategies to convert the relevant wilderness into suitable habitat for species that would otherwise be endangered, our opponent shifts the burden of proof, adding a requirement for the preservation of wilderness.

Because our adequacy conditions will be used to rule out several proposals for a conservationist ethic, and in order to prevent arguments of the sort mentioned in the last paragraph, we have to ensure that the conditions are acceptable to all those who are concerned with biodiversity conservation. In the interests of attempting to reach such a consensus, it makes sense to keep these adequacy conditions as minimal as possible. The conditions given here have been framed with this consideration in mind. Some of them have been implicitly assumed in many discussions in earlier chapters of this book. Here, the conditions will be made explicit and argued for, since they are open for debate. The final paragraph of this section will try to justify the less obvious of these conditions.

There are six such conditions:

(i) A *generality* condition: the ethic must be one that attributes value to biodiversity in general,[5] in all its complexity. It must, therefore, give credence to the idea that there is value in diversity at all levels along both the taxonomic hierarchy (of subspecies, species, genera, and all

[5] For the time being, it will be assumed that we are sufficiently clear about what general biodiversity is that we can talk about it. It is an exceptionally vague concept, as Chapter 6 (§ 6.5) will show; an attempt will be made there to clarify its definition.

higher taxa) and the structural hierarchy (of subpopulation, population, community, and so on). It must also support the protection of endangered biological phenomena, such as the annual migration of the North American monarch butterfly (*Danaus plexippus*).[6]

(ii) A *moral force* condition: the ethic must produce an obligation in us to attempt to conserve all biodiversity. Thus, the value it attributes to biodiversity must be of the sort that requires action on our part.

(iii) A *collectivity* condition: in particular, the ethic must attribute value at least to species and, ipso facto, to populations, because, in practice, these are the entities that are most often going to be the direct targets of conservation. The important point here is that attributing this value to species must not be entirely based on attributing values to individual organisms. The relevance of this point will be further discussed later, under entry (a) of the last paragraph of this section. Preferably, as implied by (i), the ethic should attribute value to other higher-level entities along both the structural and taxonomic hierarchies.

(iv) An *all-taxa* condition: the ethic must attribute value to all species (and other taxa) and not merely to higher species, "charismatic" species, or some other subset of species delimited by some criterion. For instance, a general obligation to the conservation of species does not distinguish between elephants and beetles. Conservation, in its present cultural contexts, may result in little interest in beetles compared to elephants. But, from the perspective of conserving biodiversity, an endangered beetle – for instance, the American burying beetle (*Nicrophorus americanus*) – deserves conservation action on its behalf. If the threatened Asian elephant (*Elephas maximus*), for instance, is to be given preference over the American burying beetle, such a choice requires justification on some other basis (for instance, the relative sparseness or rarity of the two genera). Initially, all species (or other units, provided that those being compared are at the same taxonomic level)[7] must be given the same potential conservation value, but endangerment, rarity, and similar criteria may then increase the conservation value of some over others (see the priority-setting condition [v] below).

[6] See Brower and Malcolm (1991); endangered biological phenomena will also be discussed in Chapter 6 (§ 6.5) in the context of the definition of "biodiversity."

[7] The restriction "same taxonomic level" is necessary in order to avoid counterintuitive comparisons, such as comparisons between an entire phylum and a single species. Clearly, protecting a phylum takes precedence over protecting a species. (If the species in question constitutes a phylum, protecting it is still protecting a phylum.)

(v) A *priority-setting* condition: the ethic must allow, if not provide a method for, the prioritization of species (or other targets) in conservation contexts. In particular, threatened or rare species, communities, and other higher-level entities should score higher than common ones. Note that without this condition, we would have no justification for many of our standard conservationist practices, including our focus on endangered species.[8] However, this condition in its strong version (demanding a prioritization method) is not as important as the three that went before. The reason for this will be discussed in entry (b) of the next paragraph.

(vi) A *non-anthropocentrism* condition: preferably, the ethic should allow us to attribute value to biological entities without reference to our most parochial human interests. Attempts to do this form the substance of section 3.3. That this condition cannot be given *any* precedence over the preceding four will be noted in point (c) below. Much of the discussion of this chapter (especially § 3.3) and the next will also underscore the difficulty of meeting this condition.

These conditions are not all independent of each other: the collectivity and all taxa conditions ([iii] and [iv]) only explicate what is intended by the generality condition (i). Moreover, not all conditions are equally important. The generality and moral force conditions ([i] and [ii]) are the most important. The generality condition (i) requires attention to all biodiversity. The moral force condition (ii) is the only one that explicitly invokes an obligation on our part to act to conserve biodiversity. The priority-setting condition (v) is less important than the first four. The non-anthropocentrism condition (vi) is the least important of them all, and perhaps should not be introduced. The ethic defended in the next chapter will not satisfy the non-anthropocentrism condition. It is introduced in this discussion only because there is wide support for a non-anthropocentric biocentrism among many biodiversity conservationists (see § 3.4 and entry [c] of the next paragraph).

Conditions (i), (ii), and (iii), of generality, moral force, and all taxa, are probably uncontroversial and do not require further defense beyond what was said when they were introduced. Three additional points need to be made to defend as well as to limit the conditions of collectivity (iii), of

[8] Moreover, the place prioritization procedures that will be discussed in Chapter 6 (§ 6.2) implicitly assume that such a prioritization of the relevant (true) surrogates for biodiversity has taken place.

priority setting (v), and of non-anthropocentrism (vi), respectively:

(a) Returning to an example introduced at the beginning of Chapter 1, for the survival of endangered sea turtles (Chelonia sp.), stray dogs may have to be removed from regions surrounding the few and very specific beaches around the world (for instance, those in Orissa in India or at Tortuguero in Costa Rica) where these turtles habitually nest. Dogs excavate and eat turtle eggs, which incubate while buried in the sand. In most situations, the only practical method for removing these predators is by eradicating them.[9] A policy of culling, whether it be of dogs, as in this case; or of African elephants (*Loxodonta africana*) in overpopulated southern African reserves;[10] or of feral goats that are threatening the fragile indigenous flora of many islands; or of cats destroying endemic and indigenous fauna through predation,[11] is almost inevitable in many conservation contexts. An ethic that is supposed to underpin conservationist efforts must permit this choice. This is why the collectivity condition, which asserts the value of species independent of that of individual organisms, was introduced. This also puts any prospective proactive conservationist ethic potentially at odds with the ethics that provide the basis for "animal welfare" and "animal liberation" movements (see § 3.5). Animal ethics cannot easily permit the sacrifice of individual sentient animals in order to meet conservationist goals. This point will be of critical importance in the assessment of some of the arguments for attributing intrinsic value to species and other nonhuman biological entities (in § 3.3).

(b) It is important that we be able to prioritize species (and other biodiversity targets) when places are targeted for biodiversity conservation, because, unfortunately, human and other material resources are likely to be severely limited in almost any conservation context. This is the role of the priority-setting condition (iv). Moreover, there may even be contexts in which we would have to consider eradicating (or at least allowing the extinction of) rare or threatened species in order to protect those with even higher priority. However, a rule that requires a conservationist ethic to provide a definite method for prioritizing every species in

9 Of necessity, this is already practiced in Costa Rica in the Tortuguero region. For the problem of dog predation there, see Fowler (1979).

10 See, for instance, <http://www.enn.com/news/enn-stories/2000/09/09142000/elephant˙31448.asp?P=1>; culling may be avoidable at the higher economic cost of sterilizing elephants. This will be discussed at the end of the chapter (see § 3.5).

11 On both goats and cats, see Ebenhard (1988).

every context would be too strong. Too many highly context-dependent factors may enter into this prioritization for them to be usefully inserted into a general conservationist ethic. Therefore, a strong version of the priority-setting condition (v) that would *require* the provision of such a definite method should not be imposed with the same status as the generality, moral force, and collectivity conditions (that is, [i], [ii], and [iii]). However, a permissive version – that a conservationist ethic *permit* such a prioritization – is just as important as those three conditions.

(c) Biodiversity conservationists tend to have an appreciation of living nature that goes well beyond recognizing only the value that other living forms have for humans.[12] This appreciation often provides the motivation for those who devote their livelihoods to conservation. Though no quantitative sociological data seem to be available, it is more than likely that such individuals have been critically important for the survival of many species, especially during the last half-century of activist conservation. Because of its ability to motivate activist conservation, it would be desirable if this aspect of conservation – an appreciation of the value of nature completely independent of human concerns – were captured in a conservationist ethic. The arguments that try to attribute intrinsic value to nonhuman biological entities (in § 3.3) attempt to do just that. However, from a strictly logical point of view, the non-anthropocentrism condition (vi) is not required in order to account for the motives of such conservationists. Appreciation of, or even reverence for, nonhuman living forms still has a human referent, the individual who does the appreciating or revering. As was pointed out at the beginning of Chapter 2, even regarding some entity as a resource does not necessarily demean it. Thus, an ethic independent of all human concerns is not logically necessary. Consequently, the non-anthropocentrism condition (vi) is clearly not on a par with the others. This is fortunate, because, as the discussion of intrinsic value will show (in § 3.3), the non-anthropocentrism condition cannot be fully satisfied.

3.2. TWO CONCEPTS OF INTRINSIC VALUE

One common strategy that is used to satisfy the non-anthropocentrism condition ([vi] of the last section [§ 3.1]) is to claim that species – or some

[12] See the detailed, empirically driven discussion of Kellert (1996).

other nonhuman taxonomic or structural unit – possess "intrinsic value."[13] Few terms have been used more sloppily in the literature of environmental ethics, and an attempt will be made here to break with that practice. First, there is a distinction to be made between claiming that (i) some quality, such as pleasure or knowledge, has intrinsic value and claiming that (ii) some entity, such as a human individual or a nonhuman species, has intrinsic value. Standard (that is, human) ethics usually discusses claims of type (i): hedonism, for instance, is the claim that all and only pleasure has positive intrinsic value, and that all and only pain has negative intrinsic value. By contrast, environmental ethics often discusses claims of type (ii), for instance, whether species or ecosystems have intrinsic value. This book will follow the practice of environmental ethics. However, it will be assumed that claims of type (ii) make sense only insofar as they are at least implicitly grounded on claims of type (i): an entity can have intrinsic value only because of some intrinsically valuable quality it possesses. Thus, in the next section, arguments for attributing intrinsic value will be evaluated on the basis of the features they posit for the entities presumed to be carriers of intrinsic value. Second, intrinsic value can be positive or negative: recall the case of pleasure and pain. In what follows, "has intrinsic value" will be shorthand for "has positive intrinsic value" (and similarly for related locutions).

Beyond these clarifications, there are at least two important senses of "intrinsic," usually conflated in environmental ethics, that should also be distinguished:[14]

(i) intrinsic_1: an entity has value only because of its own (intrinsic_1) properties, including properties between its constituent parts, and not because

[13] O'Neill (2001) and others have pointed out that the move to attribute intrinsic value to nonhuman natural entities often, though not always, involves a meta-ethical commitment to ethical realism that usually and incorrectly supposes that such a commitment is necessary for any environmental ethic. However, as O'Neill (2001) also points out, no normative environmental ethic requires a commitment to a particular meta-ethical stance. The positions advocated in this book are neutral with respect to the dispute between ethical nonrealism and realism (as they are with respect to other meta-ethical disputes).

[14] See Korsgaard (1983) and O'Neill (1992, 2001); O'Neill (1992) also distinguishes a third sense of intrinsic value, "objective value," the attribution of which, as he correctly notes, constitutes a meta-ethical rather than an ethical claim. O'Neill (2001) further distinguishes between two types of what is called intrinsic_2 value in the text – noninstrumental value, as explicated here, and a type of intrinsic value that confers ethical standing to entities (which is O'Neill's interpretation of Kantian ethics). However, as the discussion of the next chapter will show, attribution of ethical standing to entities does not require an attribution of intrinsic value in any sense. Consequently, the last-mentioned distinction will not be adopted here.

of any property of some other entity, including whatever relations that may hold between it and that other entity. Irreducibly intrinsic$_1$ properties of an entity are those that can be defined without any reference to other entities. In this sense, they are (externally) nonrelational. Examples include the property of being bilaterally symmetric or, more relevant in the context of biodiversity conservation, being endangered.[15] In sharp contrast are relational properties, for instance, the property of being rare or being closely evolutionarily related to some long-extinct taxon.[16] Intrinsic$_1$ is to be contrasted with extrinsic. Relations with other entities are extrinsic relations. It is obviously possible for an entity to have both intrinsic$_1$ and extrinsic properties; that is, the two are not mutually exclusive.

(ii) intrinsic$_2$: whatever value an entity has is irrespective of whatever value it may have as a means to the ends of any other entity (typically humans, in the context of biodiversity conservation). Intrinsic$_2$ value is primarily to be contrasted with "instrumental" value: the value that some entity has because of its utility in producing features that are more directly valuable to others; if these others are human, then these values are anthropocentric.[17] However, from a logical point of view, such a limitation to human interests is not required. Consider, for instance, Leopold's famous "land ethic": "[a] thing is right when it tends to preserve the integrity, stability and beauty of the biotic community. It is wrong when it tends to do otherwise."[18] This claim is often taken to be an assertion of intrinsic$_2$ value (see § 3.3). However, what it actually asserts is that the "thing" referred to is "right" if it has instrumental value for a biotic community. (Leopold assumes, without argument, that communities have specifiable ends.) The "thing" in question is either an action directed at, or otherwise involving, some component of the biotic community (which is the usual reading), or such a component itself. The definition leaves open, and Leopold's discussion does suggest, the possibility that the community as a whole does have intrinsic$_2$ value. This interpretation of Leopold is supported by another typical

[15] The fact that external factors may be responsible for making an entity endangered, or that this status can be changed by manipulating these external factors, has no bearing on the fact that being endangered is a property of the entity alone.

[16] Even rarity is problematic, as will be discussed later in the text.

[17] Anthropocentric values will be discussed in greater detail in the next chapter.

[18] See Leopold (1949), pp. 224–225, which is among the most influential positions in environmental ethics. For more on Leopold's influence, see Callicott (1989).

quotation from Leopold: "If the land mechanism as a whole is good, then every part is good, whether we understand it or not."[19]

Note that, according to these definitions, entities can have both $intrinsic_1$ and $intrinsic_2$ value as well as only one or the other (or, obviously, neither). However, if an entity has *only* non-$intrinsic_2$ value, it cannot have $intrinsic_1$ value. For both meanings of "intrinsic," the ultimate conclusion desired by advocates of intrinsic value is that, if a species (or other unit) has intrinsic value, then, simply because it is a repository of intrinsic value,[20] it has to be treated in such a way that no harm comes to it. However, almost all advocates of such attributions of intrinsic value interpret it as $intrinsic_2$ value, that is, they want to conclude that all components of biodiversity deserve protection irrespective of whether they contribute any value, directly or indirectly, to humans.

The idea that much of biodiversity has $intrinsic_1$ value in one mitigated sense should not be too controversial. For instance, we often value a different species (or some other biotic entity) because it presents us with some set of properties that are different from those we are familiar with. Now even though this difference may initially appear to be a relational property, it is often reducible to the properties that the species (or other entity) has

[19] See Leopold (1950), p. 146. Leopold goes on: "If the biota, in the course of aeons, has built something we like but do not understand, then who but a fool would discard seemingly useless parts? To keep every cog and wheel is the first precaution of intelligent tinkering." This argument provides much better support for conserving every feature of biodiversity than the excessive rhetoric of the rivet argument (recall the discussion of Chapter 1, § 1.3).

[20] This does not quite say, but comes perilously close to, saying, that a species or some other target of biodiversity conservation must be treated with respect simply because it exists. While the last position (attributing "existence value") has some intuitive appeal – and some adherents among environmental ethicists – it is practically useless as an ethic to found conservation practices. There are at least two problems from which it suffers. (i) We may choose to value everything that exists and try to avoid harming anything. This still permits a kind of quietism, reminiscent of certain ancient Chinese and Indian worldviews. However, it will not permit – and certainly does not require – the interventionist conservation practices (in the extreme case, the culling of unwanted animals) that biological conservation requires. (However, treating other species as repositories of intrinsic value does not logically lead to an all-encompassing quietism.) (ii) Arguing for conservation from an assumption of existence value is tantamount to saying that we should conserve anything that exists. It should not go unnoticed that such an argument underlies many politically fashionable positions, such as many cultural relativist programs for the preservation of cultural diversity. However, this same argument can be deployed to argue for the conservation of slavery, racism, patriarchy, sexism, and miscellaneous other social and other practices that, presumably, most of us find somewhat objectionable. For a different objection to using existence value in this context, as suggested by economists, one that argues that it is ultimately anthropocentric insofar as it is based on individual preferences, see Wood (2000), p. 200, n. 18.

by itself, that is, its intrinsic$_1$ properties. For instance, the Australian gastric brooding frogs (Rheobatrachus sp.) mentioned in Chapter 2 (§ 2.1) were fascinating because of their peculiar breeding habits; their presumed extinction is a serious loss of biodiversity. *Rheobatrachus silus* was discovered in the Connondale Ranges of Queensland in 1973.[21] The female swallowed its fertilized eggs, which developed into young adults in its stomach.[22] It turned off its digestive mechanisms during this six-week period and did not eat until the offspring finally emerged through its mouth. *R. silus* initially appeared valuable to us because of its strange behavior. Being "strange" is a relational property: an entity is "strange" because it differs from those that are deemed "normal" on descriptive grounds. But the description given here required no reference to these other entities. Even though we may find *R. silus* strange because its behavior is unlike ours, nevertheless, describing that behavior need make no reference to other entities: the behavior is thus intrinsic$_1$. However, it may still be argued that, short of strangeness, nothing in the behavior of *R. silus* is valuable and, consequently, that it is not worthy of conservation. In any case, by the mid-1970s it became clear that *R. silus* was endangered and thus both valuable and worthy of conservation for that intrinsic$_1$ reason alone.

The tragedy of Rheobatrachus is its disappearance so soon after its discovery. In 1976, *R. silus* was described as abundant; by the summer of 1978–79, its populations were declining. No specimen in the wild has been seen since. The last known captive specimen, a male, died in an Adelaide laboratory in 1983.[23] A second species (*R. vitellinus*), which also exhibited gastric breeding, was discovered in 1984. It disappeared in 1985.[24] These extinctions are part of what many believe to be an ongoing anuran (frog and toad) decline crisis, the disappearance of anurans throughout the world. Whether or not there is such a decline, it is clear that extinction has significantly decreased the diversity of anurans during the last two decades.

In fact, of all the properties of species (and other entities) commonly valued by biodiversity conservationists – morphological or behavioral complexity, endangerment, endemism, fragility, rarity, richness, and so on – only rarity may reasonably be viewed as not an intrinsic$_1$ property. An entity is rare only in comparison to the degrees of occurrence of other entities.[25]

[21] Liem (1973); see Sarkar (1996) for more detail.
[22] See Tyler (1983) for the only comprehensive treatment of the biology of gastric brooding frogs.
[23] Tyler and Davies (1985) record the species' unexpected march to extinction.
[24] See McDonald (1990) for a detailed record of the second extinction.
[25] O'Neill (1993) thus takes rarity to be a property that is irreducible to intrinsic$_1$ properties.

However, it can even be argued that rarity can also be defined nonrelationally, using geographical range as the relevant criterion: the rarity of an entity is the inverse of the frequency with which it occurs in a set of places. Thus there is a sense in which even rarity, defined in this way, can be viewed as an intrinsic$_1$ property. Besides geographical range, rarity can also be defined using two other criteria: habitat specificity – the more specific an entity's habitat is, the rarer it is; and local population size – the rarity of an entity is the inverse of its local population size.[26] Suppose that all three criteria are used, and that the satisfaction of any one up to some threshold is sufficient to declare an entity rare. Then there are seven notions of rarity, each of which is nonrelational.

Returning to intrinsic values, the trouble with an attribution of intrinsic$_1$ value to biodiversity is that it does not give us the resources that are especially useful in generating moral obligations. Usually, in defending the claim that person *X* is morally obligated to preserve or defend entity *Y*, we cite a relation between *X* and *Y* that will underwrite the obligation. Perhaps *X* has promised to protect *Y*, or perhaps *X* stands in some kind of debt to *Y*. Since intrinsic$_1$ values are nonrelational, they cannot connect persons to biodiversity in the familiar way that generates many of our moral obligations. Indeed, when we look at some of the properties that are supposed to confer intrinsic$_1$ value, it is unclear how they could generate ethical obligations. Mere morphological complexity entails no ethical imperative. Peculiar behaviors may well be pathological, something to be avoided at all costs rather than carefully preserved and nurtured. We are happy that serial murderers are rare; we would not bemoan their extinction. In this sense, an appeal only to intrinsic$_1$ value falls afoul of the moral force adequacy condition (ii) introduced in the last section. In fact, what this argument calls into question is whether intrinsic$_1$ values confer *value* at all, since their possession fails to carry normative implications with it.

One way out of this problem is to argue that *we* value biodiversity only because of its intrinsic$_1$ properties, but now *we* have entered into the picture, and any property that confers biodiversity its value will be at least in part relational and not purely intrinsic$_1$. It is, therefore, not surprising that those who would base biodiversity conservation upon an attribution of intrinsic value to biodiversity usually follow the strategy of attributing intrinsic$_2$ value to it. This strategy will be discussed next. For the sake of

[26] See Rabinowitz et al. (1986) for an extended discussion of concepts of rarity and their relevance to biodiversity conservation.

expository convenience, for the rest of this book, unless explicitly indicated otherwise, "intrinsic" will be used only in the sense of "intrinsic_2."

3.3. ARGUMENTS FOR INTRINSIC VALUE

The strategy of attributing intrinsic (intrinsic_2) value to nonhuman species (or other biological entities) extends to those entities one of the more venerable traditions of founding ordinary (that is, human) ethics. Each human being is presumed to have the same ethical standing: every individual must be respected as an individual on a par with any other. This equality of ethical standing can be argued for in a variety of ways. For Kant, for instance, no human being can be treated merely as a means to an end; each person must always also be treated as an end-in-itself. This leads to a modified version of the biblical admonition, the so-called Golden Rule, that you should never do to others what you would not have them do to you. Thus human beings cannot be treated merely as bearers of instrumental value. They must also bear intrinsic value. Moreover, the attribution of intrinsic value brings with it enough normative force to satisfy the moral force adequacy condition ([ii] of Section 3.1). Historically, many reasons have been given why all human beings have intrinsic value: they all have the capacity for reason, they are sentient or conscious agents, and so on. This is not the only way to found ethics, but it is a respectable and influential tradition. What is critical about the attribution of intrinsic value in this fashion is that it naturally carries an obligation to act through a principle that may be regarded as one of symmetry or reciprocity: since other entities have intrinsic value in exactly the same way we do as individuals, these other entities are entitled to the same consideration from us that we have for ourselves. If features of biodiversity have intrinsic value in the same way we do, we must try to protect them in the same way we try to protect ourselves. They are on a par with us.

Our question is whether we can sensibly attribute intrinsic value to species (in particular, to nonhuman species) and other units in the structural and taxonomic biological hierarchies. Many conservation biologists – and some environmental ethicists – believe that we can, and attempt to found biodiversity conservation on that basis. For instance, Meffe and Carroll, in their seminal textbook of conservation biology,[27] after considering arguments for and against such an attribution of intrinsic value, endorse it on the

[27] Meffe and Carroll (1994); the significance of this book in the history of conservation biology will be noted in Chapter 6.

grounds that it will lead to a better attitude on our part in our interactions with other living forms. They quote Fox with approval:

> "Recognizing the intrinsic value of the non-human world has a dramatic effect upon the framework of environmental debate and decision-making. If the non-human world is only considered to be instrumentally valuable then people are permitted to use and otherwise interfere with any aspect of it for whatever reasons they wish.[28] If anyone objects to such interference then, within this framework of reference, the onus is clearly on the person who objects to justify why it is more useful to humans to leave that aspect of the world alone. If, however, the non-human world is considered to be intrinsically valuable then the onus shifts to the person who wants to interfere with it to justify why they [*sic*] should be allowed to do so."[29]

This is intuitively an appealing argument, though it is far from obvious that the transition from instrumental to intrinsic value is required for (or necessitates) the claimed shift of obligation. For instance, an instrumentalist can reasonably claim that a symmetric evaluation of utilities is what is required in the first place, and that the switch from instrumental to intrinsic value makes no difference. However, the most important conclusion that it implies, that a new attitude toward the nonhuman world, an attitude different from the one we customarily display, would better safeguard biodiversity and environmental health, is almost certainly correct. It will be a major virtue of any environmental ethic that leads to this conclusion.

However, philosophically, at the very least, even besides the one problem already noted, it is a baffling argument. Normally, attributions of intrinsic value occur in contexts in which we look for ethical foundations that do not depend on consequences. Otherwise, we could rely on mere instrumental value. Here we are supposed to attribute intrinsic value because of the consequences of that attribution. But what if we decide to make such an attribution and then discover that prior reasons make it incorrect to do so? Before this argument is accepted, we must ensure that we even understand what it means to attribute intrinsic value to anything other than human individuals.

[28] That this claim is manifestly false will emerge from the discussions of the next chapter.

[29] Fox (1993), as quoted by Meffe and Carroll (1994), p. 29. In sharp contrast, Wood (2000) argues that attributing intrinsic value to nonhuman entities would be politically inexpedient (see also Aiken 1992). This is one of his reasons for adopting an anthropocentrism that is in many ways similar to the one adopted in this book. (Note that the soundness of Wood's prudential argument is not being assumed here.)

In attempts to attribute intrinsic value to nonhuman species (and other higher-level biological features) in conservationist writing, two rhetorical strategies are standard. The first, more common among biologists than among philosophers, is an implicit appeal to intuition: assert that other species have intrinsic value and expect agreement.[30] Since many of us already value biodiversity and enjoy the existence of other species (and other components of biodiversity), we are tempted by this move. But this strategy works only when we are preaching to the converted; in other contexts, it is useless. This appeal to intuition is sometimes bolstered by some version of arguments (i), (ii), or (v) below (that is, the arguments from religious authority, expansion of the moral circle, and holistic rationalism). Those arguments are defective, as will be shown during their subsequent discussion, but their introduction nevertheless adds some rhetorical force to the appeal and may win some new converts. The second strategy is firmly based on the arguments to be given here but, typically, also hints that those who disagree with such attributions of intrinsic value do not appreciate the value of biodiversity. This move probably also adds rhetorical force to this strategy. But it has no intellectual merit.

It is time, then, to turn to sustained explicit attempts to defend the attribution of intrinsic value to nonhuman entities by embedding such attributions in an adequate system of values. There is a wide variety of such attempts. The seven – though the first is not really an argument – that are analyzed here are intended to reflect the full spectrum of arguments that have so far been offered. However, the list is not a comprehensive one.[31] Each argument will be presented and then criticized, in part using the adequacy criteria of the previous section:

(i) *Religious authority*: this position, which is irrelevant to philosophy, attempts to provide a religious justification for the attribution of intrinsic value of other species. For instance, the biologist David Ehrenfeld adopts a reading of the Old Testament ("J-theism") that identifies different authors of the text, the oldest of which is supposed to have provided an injunction against harming other species.[32] More

[30] For a philosophical example, see Naess and Sessions (1995).

[31] These choices have been guided largely by the very helpful discussions in Callicott (1986) and Norton (1987).

[32] See Ehrenfeld (1976); this "argument" also implicitly makes the rather dubious suggestion that the oldest is the most authentic and, therefore, most important. Should we preserve discrimination against women simply because in most, though not in all (in particular, not in matriarchal) societies such discrimination has an ancient legacy? In Ehrenfeld's biblical context, this may make some sense: with the passing of time there is everywhere a fall from grace. Recall that this

straightforwardly, many authors attempt to found such injunctions on Jainism and other Indian or Chinese religions. These positions need not detain us here; they are based on texts that are sacred only to some and have no particular normative force for those to whom they are not.[33] The appeal to religious authority will be of little use with those who do not share those religious beliefs. Invoking religious reasons may even be rhetorically inopportune in conservation contexts, if those to whom conservationist concerns must be directed do not share, or perhaps even violently disagree with, the religion that is being invoked. Moreover, as mentioned in the Preface, the mere appeal to religion does not provide philosophical reasons to accept any position.

(ii) *Expansion of the moral circle*: not so very long ago – to expropriate the opening phrase of Sartre's justly famous Preface to Fanon's *The Wretched of the Earth*[34] – in Northern societies, legal rights, including that of voting, resided solely with white males who owned property. Later, these rights were extended to those white males who did not. The moral circle had expanded. Property was no longer deemed to differentiate individuals with respect to their relevant moral attributes. Even later, white females were included in the circle. Finally, people of color were deemed to be worthy of entry.[35] According to proponents of the animal welfare and animal liberation movements, this expansion can – and should – continue. It has now become time to admit animals. One possible argument for this position is that, just as we used rationality (or some other such criterion) to expand the moral circle to include all of humanity, we can use it to include animals. Moreover, if rationality provided the basis for our ethics, consistency requires that we should expand it to include all rational organisms. The principle of consistency that must be appealed to here is *maximal*: if a criterion is applied at all, it must be applied to *all* entities to which it is potentially applicable. A weaker principle – for instance, one that

assumption was called the myth of the golden age in Chapter 2 (§ 2.2). For an exactly opposite reading of the environmental message of the texts of the Judeo-Christian–Islamic tradition, see, especially, White (1967). In that tradition, White finds support for a conservationist ethic only from Saint Francis of Assisi, "the greatest radical in Christian history since Christ."

33 Of course, religious texts may contain arguments. But these are then to be judged on their own merit *as arguments*.

34 See Fanon (1963), with a preface by Sartre.

35 The order was different in different places. In the United States, for instance, males of color preceded white females.

requires only inclusion of all entities to which the criterion happens to have been applied (for whatever reason) – will not suffice. This is one problem with this argument, though not a terribly compelling one – it will not be considered any further. There are also problems as to whether a nonarbitrary boundary between rational and nonrational organisms can be drawn and whether, further, there may be normatively important distinctions between degrees of rationality. Were this argument ultimately to be endorsed, these are problems that would have to be addressed – here, since the argument will ultimately be rejected, they can be noted without further analysis.

It is unclear how many species of animals can be included using rationality as a criterion for expanding the moral circle. But instead of rationality, we could also use sentience as the criterion for expansion, especially the ability to experience pleasure and pain and, basing our human ethics on Bentham's hedonistic consequentialism,[36] expand our moral circle to include all sentient organisms. As before, in this case, because our ethics is based on the ability to feel pleasure and pain, maximal consistency requires that we so expand our moral circle to include all organisms that can feel pain.[37] If we attribute intrinsic value to other humans, we now have to attribute intrinsic value to these other organisms. Once again, there is the problem of arbitrariness and the possibility that the variability of degrees of sentience may have normative significance. Nevertheless, these problems are not as serious as those encountered in the situation when rationality is used as the criterion of demarcation: it is at least arguable that all animals that have any vestige of a nervous system should be included. If this includes some animals that are not truly sentient, such an error does not cause ethical harm, since it does not exclude any entity that should have ethical standing.[38]

[36] What is pleasurable is good; what is painful is bad. For a recent useful discussion of sentientism, see Varner (2001).

[37] The discussion in the text assumes that sentient organisms are coextensive with those that have the capacity to feel pain. This assumption may, of course, be empirically false, as Varner (1998) argues. However, this will not affect the claims being defended in the text: all that needs to be changed is that a separate argument will have to be constructed for each of the two criteria.

[38] Arguably, it may cause some harm in the sense that scarce resources could be wasted protecting those entities that should not be eligible for protection. However, such harm is likely to be trivial in most cases in the wider context of preventing much more serious harm to those that deserve protection but have traditionally been excluded from moral consideration.

For the animal welfare and liberation movements, considerations of sentience provide a rather strong argument.[39] But this argument does little to forward a conservation ethic. There are two problems. (a) It does not satisfy the collectivity adequacy condition ([iii] of section 3.1). The focus is entirely on individuals. To the extent that a species will survive if individual members of it survive, it affords an *indirect* method of preserving species; it provides no rationale for focusing on them. (b) All nonsentient (or nonrational) organisms (for instance, all plants) are excluded, even as individuals. Thus, the all-taxa adequacy condition ([iv] of section 3.1) is not even remotely satisfied. Peter Singer, perhaps the best-known philosophical proponent of the animal liberation movement, explicitly admits that these arguments do not provide any basis for the concern for species as a whole, let alone for plants and other nonsentient species.[40] The tension between a conservation ethic and animal welfare and liberation movements will be examined further in the last section of this chapter (§ 3.5).

(iii) *Conativism*: in general, "conation" refers to willing. In the context of discussions of the intrinsic value of nonhuman entities, it is usually, in one sense more narrowly, interpreted as the "will-to-live," that will being construed purely biologically. It is probably uncontroversial that, in some sense, such a purpose can be attributed to all living organisms; the biological interpretation of conation regards this purpose as a will. Thus the biological construal of conation, in another sense, makes that concept much broader than what is customarily meant by willing: it includes plants as having a *will* to live. Suppose we accept that having such a will is sufficient reason for attributing intrinsic value to an entity. This leads to worldviews such as Schweitzer's "reverence-for-life" ethic.[41] This view is more palatable than (ii). By demanding concern for every individual of every species, it provides an indirect rationale for species conservation, and it does so for all species, thus satisfying the all-taxa adequacy condition ([iv] of section 3.1). Nevertheless, for the reasons discussed earlier in (ii), it falls afoul of the collectivity

[39] Of course, if we use rationality rather than sentience as our criterion, we only get a smaller expansion of the circle. This is presumably one reason why the advocates of these movements prefer to use sentience. There is the additional problem that some humans are not rational, for instance, because of mental impairment due to disease.

[40] See, for example, Singer (1979).

[41] Schweitzer (1976). Important variants of the argument for reverence for all living forms are provided by Feinberg (1974) and Goodpaster (1978). See Taylor (1986) for a detailed discussion.

adequacy condition ([iii] of section 3.1); and it also does not satisfy even the permissive version of the priority-setting adequacy condition ([v] of section 3.1). Moreover, there is ample reason to worry about the attribution of intrinsic value on the basis of a will-to-live. It would, even if we are vegetarians, make us destroy entities that have intrinsic value so routinely that the concept of intrinsic value itself loses any obligation-inducing force, leaving the appeal to it for the conservation of biodiversity on very shaky ground.

(iv) *Identification*: the influential environmental philosopher Arne Naess attempts to found the intrinsic value of other features of the natural world on the basis of an account of our identification with them: "We tend to see ourselves in everything alive. As scientists we observe the death struggle of an insect, but as mature human beings we spontaneously also experience our own death in a way, and feel sentiments that relate to struggle, pain, and death."[42] Such "spontaneous identification" is supposed to be the source of intrinsic value. We may well be able to identify with individual *animals*. However, it is less than clear what it would even mean to identify with plants and other organisms that do not have sentient feelings. But suppose that we can do so, simply because they are alive. Even then, we cannot in the same way directly identify with entire groups of organisms, even though we may identify with every member of the group. Thus, even if we accept this argument, it does not satisfy the collectivity adequacy condition ([iii] of section 3.1): even if we accept Naess's claim, it at most lets us attribute intrinsic value to individual organisms but not to populations of organisms, and ipso facto, not to species and other higher units along either the structural or the taxonomic hierarchy. The failure to satisfy the collectivity adequacy condition results in a similar failure to satisfy the all-taxa and priority-setting adequacy conditions ([iv] and [v] of section 3.1). But most importantly, Naess's account does not do what it claims to do: provide a normative basis for attributing intrinsic value to other species (and other biological features). At best, it only provides an *explanation* of the attribution of intrinsic value to other individuals provided that we already have made such an attribution. It provides no normative justification for such an attribution. It may show why we attribute intrinsic value to nonhuman individuals: because of identification. But it gives no reason why we *should* forge such an identification.

[42] Naess (1986), p. 506. No further argument is given in defense of this claim.

(v) *Holistic rationalism*:[43] this is a very popular interpretation of Leopold's influential "land ethic," which was introduced in section 3.2. According to this interpretation, the diversity, integrity, and other similar attributes of biotic communities (and larger units) are intrinsically good and should be cherished. This puts the focus where we want it, on communities and perhaps larger units, and Leopold's land ethic is often taken as the key foundation by many conservation biologists and philosophers. The trouble is that this view simply amounts to reiterating the original problem as its solution. We set out to determine how we can say that maintaining biodiversity (ensuring the continued stability, diversity, integrity, and similar qualities of biological systems) is to be justified on the basis of attributing intrinsic value to those systems. Simply asserting that those systems have such a value is no progress at all. This is not to suggest that the land ethic cannot be justified. It is easy enough to justify it on instrumental grounds, for instance, by pointing out the harm to long-term human interests that occurs if it is ignored. Contrary to how many of his followers have interpreted him, this, in fact, is how Leopold typically argues in *A Sand Country Almanac*.[44] But this is a far cry from producing a rationale for attributing *intrinsic* values to biological communities and larger units along the ecological hierarchy.

(vi) *Interests*:[45] living organisms have interests. Conation, in its biological interpretation of will-to-live, gives rise to some such interests, but so do other features that are universal to living forms, especially what we could by analogy to will-to-live call "will-to-reproduce." Ever since Aristotle, philosophers (and others) have often found it straightforward to talk about what it is for biological entities to flourish or be harmed, without reference to human interests or values.[46] Darwinian reproductive fitness is one obvious measure of such organismal interest. That organisms have such interests is plausible enough that the legal scholar Christopher D. Stone once urged legal standing for trees and other natural objects such as mountains, rivers, and lakes.[47] This

[43] The terminology is that of Callicott (1986), p. 148.

[44] Leopold (1949); see Norton (1991) for a balanced reading of Leopold's legacy.

[45] This important argument goes back to von Wright (1963).

[46] See, for instance, Taylor (1986) and Atfield (1987).

[47] See Stone (1974) who, however, uses "needs" instead of "interests." Stone bolsters his arguments by drawing an analogy to the fact that corporations do have such standing. This is a compelling analogy, but it is quite plausible that biological entities other than humans should be granted legal rights without attributing intrinsic value to them. Such a position, as an argument for general

argument permits an attempt to attribute intrinsic value in the same way as in (iii) and thus suffers from the problems mentioned there. However, unlike conation (interpreted as will-to-live), one might attempt to attribute interests to species or even communities or other higher-level entities. We can, without too much difficulty, often say what it is for a species or a community to flourish: it is expanding its range, it shows no sign of disease or impending extinction, and so on. The trouble is that it becomes much harder to specify these interests as we climb either the taxonomic or ecological hierarchy, especially when we consider evolutionary time scales.[48] Species become extinct naturally; some speciate by branching, others by transformation. Communities change radically; some species "benefit" through the extinction of other species while ecological succession is taking place. If we accept all the changes that higher taxa undergo as being in their interests, we have no rationale for species preservation, that is, we fail to satisfy the collectivity adequacy condition ([iii] of section 3.1), though in a different way than in the previous arguments. We also have no handle on the all-taxa and priority-setting adequacy conditions ([iv] and [v] of section 3.1). If we want to identify a certain subset of these changes that serve conservation, we must be able to do so using natural criteria that are characterized independently of these conservationist goals. No one has proposed such a subset, and it is doubtful that such a project can be successfully executed.

(vii) *Bio-empathy*: Callicott extends the identification argument in an interesting way by attempting to found intrinsic value on a naturalistic basis, drawing on a version of sociobiology. Individuals have their own specific selfish goals, but they also have goals that bind them to related individuals, and onward to larger entities, ultimately to all of life. All of this is supposed to be driven by natural selection (at levels higher than that of the individual).[49] This is supposed to lead to empathy with an ever-widening circle of living entities. Just as the intrinsic value of

biodiversity conservation, seems quite promising and is very different from the one advocated in this book. It merits further development. However, Stone (1985) tempers many of his original claims. See Varner (1987) for further discussion.

[48] Even von Wright (1963), who initiated this line of argument in its contemporary context, limited its applicability to living individuals (of any species). For him, the interests of aggregates are reducible to those of their individual members.

[49] See Callicott (1986). Wilson (1988a) develops an only slightly more sophisticated version of this view in his doctrine of "biophilia." This is Wilson's sociobiology finally raising its ugly head in a conservationist context. (On Wilson's sociobiology and the ethical and intellectual problems associated with it, see Kitcher 1985.) Wilson claims – as is often the case in human sociobiology,

other humans arises because of our empathy with – or sympathy for – them, on this account the intrinsic value of other species arises through this selection-driven expansion of our empathy. There are two serious problems with this claim: (a) It relies on what are almost entirely unsupported biological claims. There is no evidence for the forms of selection that this account requires. And (b) even if it were correct, as with Naess's identification argument, it provides an explanation of the attribution of intrinsic value to other species (and other entities) only if we have already done so. It provides no normative justification for such an attribution, no moral argument that says we should do so if we have not.[50] Thus, though it satisfies the generality, collectivity, and all-taxa adequacy conditions ([i], [iii], and [iv] of section 3.1) and, obviously, the non-anthropocentrism adequacy condition ([vi] of section 3.1), there is no good reason to accept this position. Moreover, it does not satisfy the priority-setting adequacy condition ([v] of section 3.1) at all: it does not even provide an explanation (let alone a justification) for the kind of prioritization that would allow us to give a higher priority, for instance, to conserving an endangered species than to conserving a common species that is more closely taxonomically (or genetically) related to us.

None of these arguments does what it purports to do: provide a conservationist ethic that obligates conservation of biodiversity at all levels of the taxonomic and ecological hierarchies. Arguments (ii), (iii), and (iv) are the most plausible; it is possible that (vi) can be refined to meet the objections raised in the discussion here. But these support making attributions of intrinsic value only to individuals (with the possible exception of [iii]), not to populations, and largely to individuals of sentient species. In other words, they lead to serious concern for animal welfare exactly in the direction that a conservationist ethic, which permits and may require culling, cannot easily go. This issue will be discussed further at the end of this chapter (in section 3.5). The next chapter will turn to anthropocentric value and try to find perhaps a possibly less emotionally satisfying but philosophically more secure rationale for biodiversity conservation.

Finally, it is important to note a limitation of the critique of attributions of intrinsic value presented in this section. Each of the seven arguments

with no empirical evidence – that there is a genetic basis for our love of biodiversity. Given the wanton destruction of biodiversity we see around us, these genes must be rather impotent.

[50] The last point has been forcefully argued by Norton (1987), pp. 172–175, who also provides other arguments against Callicott's position.

for attributing intrinsic value was examined and found wanting. However, this leaves open the possibility that some other successful argument for attributing intrinsic value can yet be found. The discussion of this section should thus not be regarded as definitive with respect to the rejection of the use of attributions of intrinsic value to nonhuman entities to provide a normative foundation for biodiversity conservation.

3.4. BIOCENTRISM AND DEEP ECOLOGY

An attribution of intrinsic value to other species (and to other biological entities) forms part of an ideology usually called "biocentrism"[51] (but sometimes also "ecocentrism"),[52] to be contrasted with "anthropocentrism" or "humanism." The latter two positions, according to extreme self-styled biocentrists, suffer from a human chauvinism, in that they found normative systems on human interests. For some, anthropocentrists are guilty of "speciesism."[53] As is far too often the case, rhetoric is a useful recourse when there is no good argument. Unfortunately for biocentrists, attributing intrinsic value to all biological entities at all levels of organization, as was shown in the last section, founders under philosophical scrutiny. If that really does leave us as "speciesists," there is little that can be done about it.

Biocentrism is often found in association with the ideology of "deep ecology," which has been endorsed by many conservation biologists. Deep ecology began its career among Norwegian environmentalists. In the Anglophone world, its founding document is a 1973 piece by Arne Naess, "The Shallow and the Deep, Long-Range Ecology Movement."[54] Naess rejected what he called shallow ecology because its central objectives were to combat pollution and to conserve resources in order to ensure "the health and affluence of people in the developed countries."[55] Instead, he proposed

[51] See, for example, Taylor (1986) for a sophisticated presentation of biocentrism in general. The criticisms of deep ecology made in this section do not address the version of biocentrism discussed in Taylor's book.

[52] See, for example, Rolston (1988). However, Wenz (1988) and others distinguish between biocentrism and ecocentrism, using as a criterion the fact that the former attributes value to individuals of nonhuman species, whereas the latter attributes value to collective or composite entities such as species. This more restricted sense of biocentrism is not being used here.

[53] See Singer (1975). Speciesism is supposed to be as blameworthy as racism and sexism. The term "speciesism" was apparently coined by Richard Ryder in 1970. See <http://en.wikipedia.org/wiki/Speciesism>.

[54] Naess (1995). For a philosophically sophisticated analysis of Naess's piece, see Mathews (2001).

[55] Naess ([1973] 1995), p. 3.

a set of principles that endorsed not only "biospherical egalitarianism" and "diversity and symbiosis" but also traditional sociopolitical agendas of the Left such as an "anti-class posture" and "local autonomy and decentralization."[56] Underlying these political aims was a relational ontology: individuals at every level of material and ecological organization were defined only with respect to their relations to all other entities in the universe. From this assumption it is supposed to follow that, in the ecological hierarchy, all units had the same moral worth. Nothing privileged humans, but Naess did not, at this stage, take the step of attributing intrinsic value to nonhuman entities. Philosophically, Naess's position is respectable only to the extent that we can accept his argument from relational definition to moral worth. And that argument is somewhat odd: every human being has a deep relation to oxygen, and *Homo sapiens* has been umbilically connected to it since the beginning of its existence. Should we, therefore, decry the use of oxygen to burn hydrogen as a way to produce water? In the political arena, the deep ecological agenda of 1973 was concerned equally with eradicating poverty in the South and preserving the general long-term health of the biosphere. Naess implicitly suggested that the former is required for the latter, though, explicitly, he claimed only that: "[t]he principles of ecological egalitarianism and of symbiosis support the . . . anti-class posture."[57] As a political manifesto, there is little in that piece that should provoke criticism.

Unfortunately, deep ecology lost its social conscience during its transplantation to North America. An attribution of intrinsic value replaced the argument from a relational ontology for respect for the nonhuman world. The basic principles of deep ecology, as Naess and George Sessions elaborated them in 1984, make no mention of human poverty or inequality. Rather,

> 1. The well-being and flourishing of human and nonhuman Life on Earth have value in themselves (synonyms: intrinsic value, inherent value). These values are independent of the usefulness of the nonhuman world for human purposes.
> 2. Richness and diversity of life forms contribute to the realization of those values and are also values in themselves.
> 3. Humans have no right to reduce this richness and diversity except to satisfy vital needs.

[56] Ibid., p. 4, p. 5, p. 6.
[57] Ibid., p. 5.

4. The flourishing of human life and cultures is compatible with a substantial decrease of the human population. The flourishing of nonhuman life requires a decrease.
5. Present human interference with the nonhuman world is excessive, and the situation is rapidly worsening.
6. Policies must therefore be changed. These policies affect basic economic, technological, and ideological structures. The resulting state of affairs will be deeply different from the present.
7. The ideological change is mainly that of appreciating *life quality* (dwelling in situations of inherent value) rather than adhering to an increasingly higher standard of living. There will be a profound awareness of the difference between big and great.
8. Those who subscribe to the foregoing points have an obligation directly or indirectly to try to implement the necessary changes.[58]

Thus an attribution of intrinsic value to all entities at all levels of the ecological hierarchy entered into the new deep ecology at its very foundations; other than this feature, there is scant attention to the philosophical foundations of the position presented in this platform. Wilderness preservation soon became central to deep ecology because it underscores noninterference with nature.

Compared to most deep ecologists, Naess and Sessions have only a rather mild distaste for humans, and they put a welcome, if only a slight, emphasis on the desirability of controlling overconsumption (item [7] of the platform). But the sparse content of the assumptions of this platform, and the ambiguity of what "deep" means in this context, has led deep ecology to be interpreted differently by its many adherents. Much more typical than Naess and Sessions are those mentioned in Chapter 2 (§ 2.2), who view humanism as essentially flawed, perhaps even evil. The political scientist Luc Ferry has pointed out that European versions of deep ecology – at least in France and Germany – also endorse an extreme antihumanism.[59] Meanwhile, ecofeminists and social ecologists are among those who have provided very important political critiques of deep ecology.[60] The troubling sociopolitical agendas associated with deep ecology in North America will not

58 Naess and Sessions (1995), pp. 49–50.
59 Ferry (1995), pp. 70–78.
60 For ecofeminism, see, for example, Salleh (1984). Ecofeminism is unfortunately receiving short shrift in this book, but only because it very rarely directly grapples with issues of biodiversity conservation. See, however, Shiva (1991), though whether this piece should be considered a work in ecofeminism, rather than in social ecology, remains contentious. For more on social ecology, see Guha (1994).

be examined in any detail here. Deep ecology was endorsed by Earth First! in its original misanthropic, messianic phase.[61] Some Earth First!ers endorsed what amounts to an eugenic agenda, with environmentalists, rather than particular races or classes, as the group to be privileged for reproduction.[62] One particularly unfortunate point came in 1986, when Dave Foreman, the founder of Earth First!, advocated anti-immigration measures specifically directed against Latin Americans and endorsed the famine in Ethiopia, both positions he later retracted.[63] There is obviously a considerable political distance from the advocacy of such positions and the environment-friendly but racist policies of the Nazis – recall the discussion of Chapter 2 (§ 2.2) – but the disturbing commonality in the negative attitude toward "undesirable" humans should not go entirely unnoticed.

What makes deep ecology attractive is exactly what makes an attribution of intrinsic values to nonhuman entities appealing: it seems to provide a better safeguard for biodiversity, and for environmental health in general, than our usual anthropocentric attitudes. It demands respect for nature and imposes obligations on us to treat it in a way that is consistent with that respect. But the failure to find an adequate rationale for such attributions of intrinsic value (see the last section) makes it impossible to formulate a credible version of deep ecology that relies on intrinsic value. Moreover, most deep ecologists couple this reliance on intrinsic value with wilderness preservationism: wilderness itself is supposed to be intrinsically valuable. The irrelevance and even occasional hostility of wilderness preservationism to biodiversity conservation pointed out in Chapter 2 (§ 2.3) also militates against any value of deep ecology for biodiversity conservation.

3.5. ANIMAL WELFARE

Animal experimentation is carried out these days in laboratories that look like fortresses. Radical animal welfare activists who are often willing to take direct action in the form of sabotage are responsible for this transformation of laboratory culture.[64] Some animal experimentation is undoubtedly of (human) medical value insofar as there is no other reasonable way to test the

[61] For an excellent history, see Lee (1995).
[62] See Noss (1984); for commentary, see Lee (1995), pp. 83–84.
[63] Foreman (1991), p. 107. Recall also the discussion of questionable political associations connected to the myth of the golden age, which is usually present in the background of biocentrism, in Chapter 2 (§ 2.2).
[64] See Horgan (1999), p. 24.

safety and efficacy of new therapies. Much – for instance, work on the brains and perceptual faculties of higher primates – is geared toward understanding biological processes analogous to those that occur in human beings but that cannot be studied on humans through experimentation because of (human) ethical considerations. It is at least arguable that much of even this research has little scientific potential, while subjecting sentient beings to significant pain and, quite often, resulting in the impairment of their faculties and even their death. There is thus a straightforward moral argument against permitting such research. But there is worse to come. Sentient animals are also used to help the satisfaction of entirely dispensable human desires. Every year, thousands of rabbits become blind in the United States because they are used to test new cosmetics.[65]

Considerations such as these motivate animal welfare activists.[66] The most common argument that is supposed to underwrite the concerns for animal welfare is the expansion of the moral circle discussed in section 3.3. But talk of *expansion* in this context is idiosyncratically Eurocentric: it uses only the peculiarly limited experience of European and neo-European societies during the last few centuries to generalize about "our" ethics. Elsewhere, concern for animal welfare and pain is of much older vintage. In the case of the Jains of India, it goes back 2,500 years. For Mahavira and other founders of Jainism, concern for the welfare of all sentient beings required no special intellectual effort and evolution. Presumably, cultural traditions of this sort have been critical to the remarkable survival of biodiversity in the Indian subcontinent, which has always been one of the most densely populated – by humans – places on Earth.

Setting aside their cultural myopia, it is hard not to sympathize with the concerns of animal welfare advocates, whether or not the actions of the more radical factions are justifiable. Moreover, to the extent that it is permissible to extend intrinsic value to nonhuman entities on the ground of sentience, such an extension supports these concerns. The trouble – in our present context – is the obvious potential conflict between animal welfare and what can only be called *intrusive* measures that may be necessary for biodiversity conservation.[67] This dispute has long been recognized among

[65] <http://www.peta/online.org/mc/facts/fsae7.html>.

[66] See Singer (1975, 2001) and Regan (1983) for powerful statements of this position in different ways. Singer relies on consequentialist arguments; Regan presents the case for animal rights.

[67] Varner (1998) has recently tried to find common ground between the animal liberation movement and biodiversity conservation while still starting from an individualist point of view. Varner's defense of hunting or culling depends on the acceptance of premises such as "more pain would be caused by letting nature take its course than by conducting carefully regulated therapeutic

biodiversity conservationists: Michael Soulé's 1985 manifesto for conservation biology,[68] which will be further discussed at the beginning of Chapter 6, explicitly drew attention to it and emphasized the differences between the conservationist and animal welfare ethics.

A few examples will drive home the practical dimensions of this conflict. Accidentally introduced brown tree snakes (also known as Philippine rat snakes, *Boiga irregularis*) have all but decimated the bird populations on the island of Guam.[69] In order to conserve birds, the snakes have to be removed. In a review of both accidentally and intentionally introduced birds and mammals in different areas of the world, Ebenhard[70] analyzed fifty-nine introductions of domestic cats (*Felis domesticus*); thirty-eight of these (or 64.4 percent) had negative effects on local prey populations, including local extinctions. This percentage is much higher than that for all predator introductions (32 percent). There is thus an obvious case for the removal of cats in many situations. Doing so by systematically slaughtering these cats would probably bother even some individuals who would not normally identify themselves with the animal welfare movement. As noted before (in § 3.1), African elephants (*Loxodonta africana*) have been culled in several southern African reserves, including the Kruger National Park of South Africa, because excessive population growth has led to a situation in which elephants destroy the habitats of other species. This culling has taken place even in a context where elephant conservation had emerged as a major continental concern in Africa. In southern India, there are similar problems with Asian elephants (*Elephas maximus*); these elephants also cause considerable damage to humans because of increasing elephant populations.[71] Culling has so far not been practiced in India, even though a good case can

hunts" (p. 106). This premise, even if accepted, provides no reason to cull dogs in order to save turtle eggs (recall the discussion of § 3.1). The only situation in which it helps is the one in which animals are being hunted because of overpopulation. Most intrusive conservationist measures that require the culling of a different species to protect another remain unsupported under Varner's proposal – there is no easy resolution to the conflict between animal liberation and biodiversity conservation. Ultimately, Varner's defense of biodiversity conservation turns out to depend on an anthropocentrism that is only partially acknowledged: only human beings have "ground projects" justifying higher or considered interests.

[68] See Soulé (1985), which is one of the founding documents of conservation biology. Callicott (1980) was among the first to recognize a tension between animal liberation and conservationist ethics. Callicott (1989), pp. 49–59, withdraws much of the inflammatory rhetoric of the earlier piece. See also Sagoff (1984).

[69] Savidge (1987); Jaffe (1994) provides a popular history.

[70] See Ebenhard (1988), which remains the most thorough such study.

[71] Sukumar (1989) provides a systematic analysis of both Asian elephant ecology and elephant–human interactions.

be made for it. In India, elephant herds can even be culled not by killing individuals but by capturing them for domestication. Unlike African elephants, Asian elephants can be domesticated.[72] (Some animal welfare theorists also find domestication ethically indefensible. However, presumably even they would concede that it is preferable to slaughter.)

Most biodiversity conservationists accept culling in such situations as an unfortunate necessity brought about by the decline in the size of habitats because of the expansion of human activities. This amounts to ignoring ethical responsibilities toward individual animals (if any), and those who find the arguments for animal welfare sound – or at least credible – cannot be happy with such a position. At the very least, from the point of view of animal welfare activists, when we have to face the prospect of removing sentient animals from a habitat, we should make every effort to cause as little pain as possible and to avoid killing to the maximal extent possible. This implies that strategies such as translocation and sterilization, whenever viable, are to be preferred over killing. In one welcome development, even in difficult cases such as that of the Kruger National Park attempts are currently being made to replace culling with sterilization.[73] However, replacing culling with sterilization will undoubtedly add significantly to the cost of conservationist measures in many of these cases. On the other hand, in some cases – such as that of the Asian elephant, which can be domesticated – culling could involve the reintroduction of elephant captures that are currently banned by law.

In conclusion, to the extent that it has forced us to accept that a genuine concern for the welfare of other sentient beings will require us to accept additional costs and to make additional efforts, the discussion of intrinsic value in this chapter has been useful, even though it has contributed little to the formulation of an adequate biodiversity conservationist ethic.

[72] See Sukumar (1994) for a development of the rationale for capturing Asian elephants for domestication as a way to minimize elephant–human conflicts.

[73] See <http://www.enn.com/news/ennstories/2000/09/09142000/elephant_31448.asp?P=2>; see also Fayrer-Hosken et al. (2000) for the design of the first potentially successful elephant contraceptive, which made this way of controlling elephant populations plausible. Pimm and van Aarde (2001) point out that the costs would be prohibitive and question the reliability of the method; for a response, see Fayrer-Hosken et al. (2001).

4

Tempered Anthropocentrism

Can we introduce value for biodiversity in such a way that (i) values remain anthropocentric but (ii) a reverence for nature is not lost? Part (ii) of this question is important because most, perhaps all, biodiversity conservationists share a reverence for nature that goes beyond appreciating its utility in the mundane sense that it delivers goods for our material consumption.[1] This chapter will try to provide an affirmative answer to this question by developing a framework for attributing values using a concept of "transformative" value originally introduced in this context in 1987 by Bryan G. Norton but surprisingly ignored in the literature since then.[2] The result is an anthropocentric defense of biodiversity conservation, but this anthropocentrism is tempered by an appreciation of the fact that biodiversity does not have the sort of human value that is routinely traded in the marketplace. It does not have cash value. The value it has is much more important. The normative account advocated here fits most easily into a consequentialist framework, though, in its discussion of the obligations for biodiversity conservation (§ 4.3), it draws on arguments that are usually associated with the deontological tradition.[3] Environmental ethics extends ethical discussion into the nonhuman realm and thereby necessarily broadens the scope of normative ethical theory – in such a context, it is unhelpful to view the different traditions within ethics (consequentialism, deontology, virtue ethics,

[1] See the sociological analysis of Kellert (1996). Recall also that this was the source of the non-anthropocentrism adequacy condition for a conservationist ethic (condition [iv] of Chapter 3, § 3.1).

[2] See Norton (1987). Even Norton has not subsequently followed through with the consequences of a transformative value-based account of a conservationist ethic, possibly because of the difficulties encountered here.

[3] Its compatibility with perfectionist ethics will also be noted here but not emphasized.

etc.) as being incompatible in principle.[4] Moreover, in the philosophically uncertain world of the ethics of biodiversity conservation, there is an obvious prudential reason to assume that normative conclusions suggested by all meta-ethical positions (or even those suggested by one and consistent with the others) should be given greater weight than those that are not. As this chapter will show, in practice it is both possible and helpful to weave a fabric from strands taken from different traditions.[5] The results are not as heterogeneous as we might initially fear; nor will the analysis avoid rigor, though there are areas in which it is still incomplete. Ultimately, though, because the normative conclusions are the ones that matter, the arguments for them are designed to be independent of meta-ethical commitments and neutral about meta-ethical disputes.

Before the argument for the use of transformative values is developed, it is worth noting a few logical points about what was achieved in the last chapter. These points should show why the positive arguments of this chapter, beyond the negative ones of the last, are required to establish a successful anthropocentric ethic of biodiversity conservation. The arguments against attributions of intrinsic value to nonhuman species (and to other biological entities at higher levels of organization) developed in the last chapter do not logically force us into accepting the idea that biodiversity must have only an anthropocentric value. In fact, there is some reason to worry that the opposite might be the case. Bearing in mind the distinction between intrinsic_1 and intrinsic_2 value, recall that the discussion of the last chapter admitted that it was reasonable to attribute intrinsic_1 value to many components of biodiversity. Unfortunately, these attributions did not seem to be of the sort that generated in us obligations to conserve biodiversity. Thus they did not provide the conservationist ethic we wanted. The contrast class of intrinsic_1 is extrinsic, and it may thus be supposed that biodiversity cannot have extrinsic value. This would be particularly troubling because anthropocentric value, as will be argued in the next paragraph, is necessarily extrinsic. However, as was pointed out in the last chapter (§ 3.2), having intrinsic_1 value does not preclude an entity from also having extrinsic value. For instance, a marvelous cave filled with stalactites and stalagmites – marvelous to *us*, for instance, because it seems to be living – clearly has both. Thus having intrinsic_1 value is compatible with biodiversity's also having extrinsic value (though this is not necessary), for instance, because of some

[4] Elliott (2001) correctly emphasizes this point.

[5] Though the result is very different, the same strategy is followed very effectively by Sylvan and Bennett (1994).

special relationship with human interests and aspirations. All that having intrinsic$_1$ value requires is that valuation on the basis of intrinsic$_1$ properties suffices to establish the value in question. We got no further in the last chapter.

Further, the subsequent arguments that were developed in the last chapter (§ 3.3) show that the standard arguments for attributing intrinsic$_2$ value to nonhuman species (and to other components of biodiversity) also do not give us the conservationist ethic we wanted. The contrast class of intrinsic$_2$ is instrumental. All instrumental value is extrinsic (the contrast class for intrinsic$_1$), since it involves a relation to some end for which the instrumentally valuable entity is a means. In most biodiversity conservation contexts, the kind of instrumental value being contrasted with intrinsic$_2$ value will be anthropocentric – an entity will be instrumentally valuable because it can bring about a state in which some human desire is satisfied or some human beings are better off. The potential for biodiversity to have anthropocentric value is consistent with what was said in the last paragraph. But we still have to show that it does have anthropocentric value. Finally, the failure of the standard arguments of the last chapter does not show that the value of biodiversity is necessarily anthropocentric. As was explicitly noted in the last chapter (§ 3.3), our discussion, even if sound, leaves open the possibility that some other argument, which we did not consider, will establish an intrinsic$_2$ or non-anthropocentric value for biodiversity. Thus, for the anthropocentric position that is being endorsed in this book, the arguments of this chapter in favor of an anthropocentric defense of biodiversity conservation will be critical. Ultimately, it is on these arguments, rather than on those of the last chapter, that the success of the tempered anthropocentric position depends.

4.1. DEMAND VALUES

Traditional economic valuation, perhaps the best-understood case of quantitative valuation, begins by assuming that the source of values is the felt preferences of individuals. Felt preferences are psychological states that determine, at least on the average, individual behavior in the marketplace. An entity has "demand value" if it can satisfy such a felt preference. The market determines what the actual demand value of an entity is. Does biodiversity get its value in this way? Certainly, those of us who are deeply committed to biodiversity conservation are willing to pay a certain amount for it. If we model this felt preference as a basis for market transactions, we will be able to attribute some demand value to biodiversity. Can we build a rationale for biodiversity conservation in this way? There have been legions of attempts to do so.[6] These have ranged from relatively plausible attempts to assess the "ecosystem services" provided by wetlands and other habitats through nutrient cycling and so on, to wishful attempts at contingent valuation – for instance, by assessing the value of national parks and biodiversity reserves by the public's "willingness to pay" for them.

The potential attractiveness of this approach lies in the fact that it falls squarely within a tradition of public policy decision making, based on quantitative economic valuation and rational choice theory, that has become the norm in North America since World War II.[7] Moreover, there are certain situations in which a purely market-based economic analysis provides good arguments for conservation. For instance, a recent report in *Nature* carried a complicated, and at least superficially thorough, analysis of the economic benefits accruing to society as a whole from leaving a habitat intact, or sustainably using it, versus converting it irreversibly.[8] For each of the five habitats analyzed – tropical forests in Malaysia and Cameroon, mangrove in Thailand, wetland in Canada, and coral reef in the Philippines – the analysis showed that purely economic criteria favored habitat nondestruction.

Nevertheless, there are ample theoretical and practical reasons to believe that these methods do not work, or at least that they do not work for all contexts of biodiversity conservation. There are at least two reasons for such pessimism about all such methods (more specific objections to "willingness-to-pay" methods will be discussed in Chapter 7, § 7.1):

(i) There is no reason to expect that every aspect of biodiversity that is biologically interesting will receive an adequately high value using methods such as those mentioned in the last paragraph. This raises the possibility that, if this is the strategy we deploy, there will be circumstances in which we will be left with no argument for the conservation of some particular component of biodiversity. For instance, the analysis mentioned in the last paragraph allowed the sustainable logging of tropical forests. This does not ensure that every species in that forest – for instance, all arthropod species – will be conserved.

[6] See, for example, Munansinghe (1993) and Barbier et al. (1995) and the references therein.

[7] Norton (1987) emphasizes this point, while also arguing that an adequate conservationist ethic must assume more than demand value for biodiversity.

[8] Balmford et al. (2002). However, this is one of very few attempts at a comprehensive assessment and may well have atypical results.

(ii) The valuations obtained are almost always open to the objection that they are little more than arbitrary and should simply be ignored during practical policy decisions.[9] At the theoretical level, do we really know what we are willing to pay to conserve a poorly known insect species in some remote corner of Peru? Or even what we are willing to pay for a piece of Amazonia? We may assign values to these under pressure from someone trying to elicit our "utility functions," but it is wishful thinking to suggest that the assignments will not be largely arbitrary. New or unknown species or habitats do not interact with our preferences in the way some tangible good with which we are familiar does. At the practical level, valuations of this sort are usually impossible to carry out in all but the simplest cases. It seems unrealistic even to suggest that they be carried out for all species, for all habitats, or for some other surrogate set for biodiversity.

This arbitrariness arises in part because of the problem of the incommensurability of different values, that is, the situation in which not all values can be ordered on the same linear scale. The line of reasoning criticized in the last two paragraphs assumes that all criteria or, equivalently, all the "values"[10] obtained using these criteria are straightforwardly commensurable. We are supposed to believe that we can order our preferences for different features of biodiversity on the same scale on which we place the cost of bread and beer. In the framework of traditional economic theory, this is the assumption that there is a single "utility function" that captures all of our values and that the rational course of action is to attempt to maximize this function. For instance, in contingent valuation, techniques such as "willingness to pay" are supposed to provide access to this function for each individual. The charge of arbitrariness is in effect a denial that such a utility function exists, that is, a claim that the function in question can be modeled only at the cost of arbitrarily assigning quantitative assessments to all our values. Genuinely incommensurable criteria (or objectives or "values") are being reduced to one measure. The problem of incommensurability is a critical one when trying to formulate specific policies for biodiversity conservation – when, for instance, the social and economic costs of a conservation plan have to be taken into account along with the biodiversity value of a place. The problem of negotiating between and synchronizing incommensurable criteria will be treated in some detail in Chapter 7 (§ 7.1).

[9] See Norton (1987), Part A, for a systematic development of the arguments of this paragraph.
[10] For a more extended discussion of this use of "value," see Chapter 7.

However, philosophically, the most important reason for resisting an attribution of demand value to biodiversity does not have to do just with the difficulty of accurately assessing and consistently assigning such values, or even with the possibility that demand values (following felt preferences) may be fickle.[11] Rather, it has to do with the more fundamental issue of whether there are some human values that should not be thought of in demand terms. Almost everyone agrees that human beings should not be traded in the marketplace: human freedom is too important to have a demand value in this sense. Most people also agree that sexual intimacy is denigrated when it is traded in the marketplace. Whether or not there are good reasons to have legal sanctions against prostitution in general, few will disagree that prostitution forced or encouraged by economic circumstance is unfortunate and undesirable. For many, the value of religious or spiritual experiences cannot entirely be captured by demand values. For those with a social conscience, access to minimal food and shelter also should not be traded in the marketplace, but, in the contemporary Northern context of triumphant liberal capitalism, these are more controversial examples. Now, note that these are all human values. In a sense, recognizing the existence of these values underscores an extreme form of the *incommensurability* problem: the values discussed in this paragraph, except for food and shelter, are radically different from those usually – and uncontroversially – traded in the marketplace. Not only are they incommensurable with the value of marketable goods and with each other, they may not even be ordered by rank. Such values are *intangible*; it logically follows that all intangible values are incommensurable with each other and with all commensurable values.[12] One human being's freedom should not be regarded as more valuable than another one's; we should not try to rank the values of sexual intimacy between different sets of individuals, and so on. The position being defended in this book is that biodiversity is similar to human freedom, or love, in the sense that it is far too important to be traded in the marketplace.

Beyond the problems associated with the practice of demand valuation, there is yet another problem that befuddles attempts to justify biodiversity conservation in this way. It will be called the "directionality problem." The discussion of this section implicitly assumes that the consequences

[11] For an interesting development of the last argument against using demand values in this context, see Krieger (1973).

[12] Different incommensurable values may each be tangible. It is just that two tangible but incommensurable values cannot be used to rank entities on the same scale.

of conserving biodiversity are desirable even if we do not know how to quantify the extent to which it is so desirable. This assumption seems so obvious that it is almost never questioned, at least in environmentalist circles. From drugs to building materials, the positive demand value of biodiversity does not seem to be reasonably questionable. But is this assumption so indubitably correct? If a biodiversity conservationist points to an example such as the Madagascar periwinkle (*Catharanthus rosea*) to bolster the case for conservation,[13] a critic can with equal ease point to a reciprocal example such as the Ebola virus. If the western African rainforest had been destroyed more systematically a generation ago, its lost biodiversity might well have included Ebola. This is a nontrivial point: there are potentially many deadly pathogens awaiting discovery in the many biologically unknown regions of the world. We do not know for sure the directionality of the consequences of conserving *all* of biodiversity. Usually we simply assume that the good will outweigh the bad, but at this level of comparison, species by species, we do not have very good reason for this belief. It may reflect an entirely unwarranted optimism on our part. Elliott Sober has argued in this context that, indeed, when we truly do not know where a course of action will take us, we have no ethical basis for going in one direction or the other.[14] Sober's response is to abandon hope for a conservationist *ethic* and to suggest that only other normative foundations, for instance, *aesthetic* ones, may provide the best resources in support of biodiversity conservation. The rest of this chapter will attempt to show that we need not give up quite so easily. Later sections (§§ 4.6–4.7) will analyze a version of the directionality problem in more detail.

4.2. TRANSFORMATIVE VALUES

Suppose that you are given a ticket to a classical concert.[15] Also suppose that you believe that you do not like classical music but you have never been exposed to it (or, at least, not in a concert setting). You know that you like blues, jazz, rock, and so on, but not classical music. Then the ticket you were given has no demand value for you: you would not have been willing to pay for it. But suppose that, on a whim, you decide to go to

[13] Recall the discussion of this example at the beginning of Chapter 2.

[14] See Sober (1986). As Section 4.4 will show, the account of transformative value that will be developed here has much in common with the invocation of an aesthetic value for biodiversity.

[15] This example is essentially due to Norton (1987), pp. 10–11.

the concert and that you enjoy it immensely. Your horizons have widened. Now you also like classical music.[16] From now, on you will be willing to pay some amount for tickets to classical concerts. Your demand values have thus been transformed by your experience at the concert. Moreover, suppose that you decide on the basis of this experience that you had been narrow-minded about presuming that you did not like certain genres of music even before being properly exposed to them. From now on, you are willing to pay some amount, though perhaps only a very small amount, for tickets to concerts of every genre of music to which you have not had prior exposure. If this is true, your demand values have been transformed quite significantly. Further, suppose that your experience has been significant enough that you begin to exhibit the same attitude toward many other forms of art, not just toward music. Your demand values have undergone an even more radical transformation.

Experiences such as the one you had in your first classical concert transformed you; because of that, the ticket you were given has "transformative" value, that is, the ability to transform demand values. It has this value in spite of having no demand value. The position advocated here is that biodiversity has value for us because of its ability to transform our demand values. There are at least two related ways, which are not mutually exclusive, in which biodiversity has such a value:

(i) *directly*, when the experience of biodiversity brings about a transformation of our demand values – immediately of features related to biodiversity but sometimes, less immediately, also of other entities. Suppose that you experience a neotropical rainforest – say, in Costa Rica – for the first time. You are overwhelmed by the majesty of the forest, the enclosed canopy above you, the green light filtering through the leaves, and the color and variety of the insects, amphibians, and reptiles around you. From this point onward, you are willing to contribute something for the protection of tropical rainforests. This is an experience that many of those who live in the North have had during the last two decades. For some, such a transformation has occurred through merely vicarious experiences of rainforests, through television, films, or photographs. All such experiences of biodiversity, vicarious or not, have immediate (direct) transformative value in the sense just indicated. Turning to what were called "less immediate" transformations, suppose that, after

[16] For the sake of this argument, it does not matter how much you like it, nor does it matter whether you like it as much as other genres of music.

your experience of the rainforest, you are outraged that the rainforests of central America have been destroyed only for the sake of creating unsustainable pastures for the production of beef. This cheap beef has decreased the price of hamburgers in the United States by about five cents.[17] You are now more than willing to pay the additional five cents for every hamburger that you consume.[18] Thus some of your demand values, but now for features not immediately associated with biodiversity, have again been directly transformed by your experience of biodiversity.

(ii) *indirectly*, when the experience of biodiversity directly leads to other developments that, in turn, lead to a transformation of the demand values of many features that may or may not themselves be associated with biodiversity. Both Wallace and Darwin came upon the theory of evolution by natural selection through their observations of the biogeographical distributions of related species. For Wallace, the crucial observations were of the distribution of plants and insects in the Amazon basin and, especially, in the Malay archipelago. These observations first led to the formulation of the "Sarawak Law," that every species comes into being in temporal and spatial contiguity with some other species closely resembling it.[19] The Sarawak Law then set the stage for the formulation of the principle of divergence of species from common ancestors, which, in turn, paved the way for the full theory of evolution by natural selection.[20] For Darwin, the crucial observations were those of the geographical variation in the beaks and other morphological features of the finches of the Galápagos. The theory of evolution is perhaps the most spectacular contribution that biodiversity has made to human knowledge. But there are many others, from Wallace's discovery of the biogeographical line named after him (which separates Indian-type and Australian-type fauna in the Malay archipelago, for instance, the

[17] See Caufield (1984), p. 109, for this figure, which is based on United States government estimates.

[18] Note that this argument is different from one for, say, the consumption of organically grown food if, in the latter case, the concern is for our own health, not for that of some other entity. If, though, the rejection of nonorganically grown food is based on a concern for the health of future generations – ensuring that soil is not "polluted" with fertilizers – then the situation is closer to the one discussed in the text. Finally, suppose that we decide not to consume a certain good because it is produced by child or slave labor. That situation is similar to the one discussed in the text.

[19] Wallace (1855). Contiguity suggests that the newer species arose by transformation of older ones; in this way, this law was an important step toward the theory of evolution.

[20] Wallace (1858) is the first full statement of the theory of evolution by natural selection. Though Darwin is believed to have arrived at the same theory earlier, it remained unpublished.

placental and marsupial mammals) to the latitudinal gradient of species richness (the steady increase in the number of native species from the poles to the equator). The promise of new insights from biodiversity seems endless. Returning to a theme treated in Chapter 2 (§ 2.1), the discussion here provides some support for the plausibility of the myth of lost futures.

The important point is that such scientific developments critically alter demand values. As will be further elaborated later, the theory of evolution has transformed human values to an unprecedented extent. But there are many less striking examples of scientific developments dependent on biodiversity that have also transformed our demand values. For those developments that even indirectly affect human technological capacities, this is an easily demonstrable point. At first glance, biodiversity studies do not appear to lead to developments of this sort, or at least not as directly as discoveries in molecular biology or in many areas of physics. However, this appearance is misleading. At the very least, knowledge of the bewildering variety of organic life has led to an understanding of the many different ways in which organisms accomplish basic functions such as locomotion, signal detection, and foraging for food. Humans can co-opt many of these for human use, thus again transforming human values. Velcro was designed in a conscious attempt to mimic the grappling hooks of some seeds.[21] Orville and Wilbur Wright are supposed to have carefully observed the flight of vultures to learn the intricacies of drag and lift.[22] Learning the exquisite details of photosynthesis and mimicking it may well be the best solution for our energy needs of the future. If achievements such as these become commonplace, we will probably reconstruct the history of science and technology very differently than we do today: James Watt's achievement in inventing the steam engine will be dwarfed by those of the biomimics of the future. Technology may soon become associated more with discoveries in biology rather than with invention and construction using inorganic materials. The demand values of many materials will change: if photosynthesis becomes the staple of energy production, fossil fuels will disappear from the marketplace, removing one of our most serious environmental problems. The invention of Velcro presumably changed the demand value of many binding

[21] See Benyus (1997), p. 4; she provides a systematic and highly readable account of such biomimicry.

[22] See Benyus (1997), p. 8, for this example, which may be apocryphal.

techniques. The advent of airplanes forever changed the demand value of oceanic travel.

Returning to Darwin, Wallace, and the theory of evolution, to a rather remarkable extent evolutionary theory has led us to recalibrate our concept of what it is to be human. Consequently, this is a case where it can plausibly be argued that a scientific development has had a transformative capacity beyond the power of money and thus beyond what can be captured by any demand value. Because of its contribution to evolutionary biology, observing and understanding biodiversity has at least indirectly transformed all human values, including demand values, more radically than any other development in human history. We no longer treat primates in the same way that we treat more distant animals. Most of us are less inclined to contribute lavishly to religious institutions because of a fear of the "Almighty." Wallace and Darwin have probably contributed more to the decline of the demand value of religion than any other individuals in history. Lesser developments than the theory of evolution are equally important. Wallace's biogeography helped to establish the theory of continental drift.[23] Even the destruction of biodiversity may lead to an understanding of the many services that natural ecosystems perform, as we see such services disappear and begin to model that process. (In this case, even the destruction of biodiversity has in a sense some positive transformative value, but only because it had so far been conserved. Since the knowledge obtained could almost certainly have been obtained in other ways, the positive transformative value of continued conservation of biodiversity far outweighs that of its destruction.)

Biodiversity is thus signally valuable because of its intellectual interest. A proactive role in its conservation is required because it is irreplaceable – extinctions are forever. Arguments from intellectual interest may seem unconvincing to those who demand immediate pragmatic virtues in the narrowest sense of immediate utility. This attitude is typical in the United States, where, for instance, the accepted legal defense of a controversial sexually explicit work of art is to claim that it has redeeming social value, not that it is, after all, a work of art. Nevertheless, the best argument for the conservation of biodiversity remains its intellectual promise. Narrow pragmatism would lead to a devaluation not just of biodiversity but of the entire scientific enterprise, if

[23] See George (1981) for a history; see Voss and Sarkar (2003) for a résumé of Wallace's biogeography.

> not of every intellectual and aesthetic aspect of human culture.[24] We should not be embarrassed to defend the importance of our intellectual interests and pursuits. All the comforts of life that are traded in the marketplace are ultimately the products of human intellectual life, of our culture, and very often – but obviously not always – of that part of our intellectual culture that we identify as science. Thus even narrow pragmatism dictates the attribution of a high value to intellectual life and science; without them, pragmatism would be useless. Janzen was correct to note that biologists have a *professional* imperative to conserve biodiversity.[25] Nonbiologists share that imperative to the extent that they value biology as a field of endeavor that should be pursued in any society that can afford it. Moreover, there is nothing that says that nonbiologists cannot share the pleasure of the knowledge that the professional pursuit of biology generates.[26] The world is interesting. We enjoy knowing about it, though some of us may cherish this knowledge more than others. Finally, for many individuals, not limited to professional scientists, scientific knowledge of the world, the sense that we are beginning to understand our surroundings, deeply affects our most basic attitudes toward all aspects of life.[27] Science has a cultural value beyond the technology that it provides.[28]

Finally, it should be emphasized again that both these ways of attributing transformative value do not necessarily deny that the value of biodiversity may be intrinsic$_1$. Both of them rely as much on the relations that hold between internal features of the components of biodiversity as they do on the particular relations that hold between these entities and external human

[24] This, as Bryan G. Norton (personal communication) has pointed out, is a nonphilosophical use of "pragmatism." Note that Norton's (1987) account of transformative value emphasizes the psychological or spiritual transformative power of biodiversity over its scientific value. However, Norton's discussion is more geared toward the preservation of nature in general (including wilderness in addition to biodiversity) than toward, specifically, biodiversity.

[25] Janzen's (1986) exhortation to biologists was discussed in detail in Chapter 2 (§ 2.1).

[26] Obviously, the same argument can be made for every other aspect of the scientific enterprise.

[27] There is an interesting and obvious connection between such an appeal to transformative values and what O'Neill (1993, 2001) calls a traditional Aristotelian account: "The flourishing of many other living things ought to be promoted because they are constitutive of our flourishing" (2001, p. 170).

[28] Thus, appreciating the transformative value of biodiversity in individual human beings can be part of a perfectionist (including a virtue-based) ethics of biodiversity conservation. Of course, perfectionism typically puts as much emphasis on nonintellectual features as on intellectual ones. See also Elliott (2001).

observers. This should not really come as a surprise – recall the discussion at the beginning of this chapter, in which it was emphasized that there is no contradiction between some entity's simultaneously having anthropocentric value and intrinsic_1 value.

4.3. OBLIGATIONS OF CONSERVATION

Even though intuitions alone contribute little to the plausibility of philosophical positions, it is striking that the invocation of transformative values accommodates a common intuition that there are entities that do not have demand values but that do, nevertheless, have anthropocentric value. Transformative values, by definition, cannot exist without demand values that are inherently human. More importantly, there is at least one sense in which transformative values are more important than demand values: the former can change the latter, whereas the latter cannot similarly change the former. There is thus an asymmetric relationship that confers more agency to the former. This is a desirable conclusion: the pursuit of human values that are not reducible to demand values is a pursuit of values that are, in this sense, presumed to be more important than the ones that can be traded in the market.[29] Transformative values satisfy this requirement and, in so doing, capture yet another intuition that motivated the initial desire to attribute intrinsic value to biodiversity, namely, that biodiversity is more important than mere utility as measured by the demand value determined by the market.

It does not necessarily follow from these considerations that all human values that are not demand values have to be, or even can be, justified on the basis of their transformative capacity. It is far from clear that, for instance, repugnance toward slavery should be viewed as a result of the ability of human freedom to transform demand values. It may plausibly be argued that such an attitude toward human freedom, regarding it as merely transformative, also denigrates freedom, though perhaps not to the same extent as treating humans as objects to be bought and sold. From many cogent ethical perspectives, freedom is a part of a normative definition of what it is to be human. From these perspectives, the value of human freedom is

[29] Note that this conclusion holds even though any transformative value is in principle incommensurable with any demand value. Even though, say, the transformative value of wilderness (see § 4.4) may be more important than the demand value of some food item, this asymmetry does not allow the rankings given to the two entities by the two criteria to be put on the same scale.

more fundamental than what can be captured by the transformative capacity of demands by some entity. The belief that sexual intimacy should not be traded in the marketplace can perhaps be somewhat better justified on the basis of the transformative capacities of sexual experiences. It is at least plausible that sexual intimacy between consenting humans serves to transform relationships and to change demand values in at least some trivial ways: for instance, through the development of sentimental attachments to material objects associated with a sexual partner. This does not preclude that there may be other ways in which sexual intimacy acquires a value that should not be regarded as merely transformative of a demand value. These other ways may be more important than this mere transformative capacity.

Turning to less clear cases, suppose that religious experiences, minimal access to food or shelter, and so on, should not be regarded as having only demand values. In contrast to the cases discussed in the last paragraph, these cases can then be more directly associated with the transformation of tangible demand values. For instance, when religious beliefs lead to charitable (or other financial) contributions, such an expenditure automatically readjusts the donors' ability to pay for many other goods and, thus, their demand values. The extent of the transformation presumably depends on the financial status of the individual making the charitable contribution. However, religious experiences may have transformative value in a much deeper way that is at least superficially comparable to the way in which scientific knowledge transforms values. They may globally alter the believer's attitude toward life. Minimal access to food and shelter straightforwardly – though rather trivially – changes most, if not all, demand values. Without such access, any rational individual is unlikely to assign much value to any other good. These other goods potentially become the focus of felt preferences only after food and shelter are already accessible. The variety of ways in which demand values may be transformed suggests that there is considerable heterogeneity in the set of transformative values. This possibility opens the door to the "boundary problem," which will be discussed later (in § 4.5).

Most importantly, unlike intrinsic$_1$ values, transformative values bring with them a natural obligation on our part to conserve those entities that bear them. As will be pointed out later, these obligations are ultimately direct obligations to other human beings, and only indirectly obligations to the components of biodiversity themselves. This is what we should expect from an anthropocentric position. One consequence of this is that, even though the invocation of transformative values accommodated some of the intuitions that led to attempts to attribute intrinsic value ($intrinsic_1$ or $intrinsic_2$ value) to many components of biodiversity, not all of those intuitions can be so accommodated.

Entities bearing transformative value impose an obligation for their conservation for two related reasons:

(i) simply because they are anthropocentric, that is, because they are human values for us as individuals. Presumably, any human ethic endorses some degree of reciprocity (or symmetry – see below), for instance, the biblical admonition (the Golden Rule) mentioned in Chapter 3 (§ 3.3). Reciprocity is a symmetric relationship that demands that one individual has some obligation toward what some other individual values, because the first individual expects the second to have some obligation toward entities that the first values. Even if you do not have the same type of transformative experience of biodiversity, you should have respect for entities that facilitate such experiences in others. This respect requires you to conserve them, at least in the weak sense of doing the least possible harm to them, because of your obligation to other human beings.[30] Note that this obligation is explicitly directed toward other human beings. Any obligation toward the entity bearing transformative value is indirect, a result of the obligation to these other human beings. Further, also note that this argument presumes that transformative values are desirable. This claim can be plausibly denied, raising the directionality problem that will be fully discussed later (see §§ 4.6–4.7). For now, suffice it to note that the argument given here attempts to show how transformative value presents an obligation for conservation of entities provided that the value in question is known to be positive.

Finally, note that the requirement of reciprocity is a much stronger one than that of mere consistency.[31] There is nothing logically inconsistent if you believe: (a) some entity *Y* is valued by some individual *X*; (b) you have no obligation to protect *Y*; but (c) *X* has an obligation to protect what you value. Nevertheless, your beliefs violate the requirement of reciprocity: by making a distinction between the type of obligation *X* has and the type that you have, they establish an illegitimate asymmetry. Why should reciprocity be accepted? Only because denying it implies a denial that, even from the narrow perspective of merely

[30] However, if you believe that no one else has any particular obligation toward any of the things that you value, this argument will carry no force – reciprocity only takes us so far and no further.

[31] As O'Neill (2001) has pointed out, spurious appeals to the power of consistency have often been the bane of writings on environmental ethics.

considering the rights and obligations individuals have *qua* individuals, every individual is not on par with every other;

(ii) because we traditionally recognize obligations toward objects that are important enough that they affect many of our other values, for instance, objects of cultural significance, such as religious or aesthetic artifacts and objects of historical interest. We recognize these obligations because we accept, at least implicitly, the power of such objects to transform our values in general.[32] In a cultural sense, they have "symbolic" value. It is possible that such traditions are no more defensible than traditions of discrimination based on gender or ethnicity.[33] Revolutionaries of every ilk question some tradition or other that societies hold sacred. When in power, revolutionaries often make a point of destroying symbols of the past. However, they generally produce their own symbols, sometimes very consciously.[34] We may disagree about what has positive transformative value, but we agree that we should preserve the things that do. Recognition of this power, and of the consequent good that these objects may contribute to human well-being (once again temporarily shelving the directionality problem), generates an interest and obligation in us for their preservation. Returning to biodiversity, we thus have an obligation to conserve it because of its transformative value. This discussion should not be taken to suggest that biodiversity has – or has only – symbolic value. The transformative value of biodiversity that was discussed in the last section (§ 4.2) did not refer to symbolic importance; rather, the transformations described there were taken to have been literally produced by experiences of biodiversity.

4.4. WILDERNESS AND AESTHETIC APPRECIATION

The discussion of the last section underscored the point that biodiversity is not the only entity that has transformative value. Two other entities are particularly important in the context of environmental protection, though for rather different reasons – wildernesses and aesthetically interesting objects. These merit some further discussion:

[32] Transformations can, of course, be negative: this is a new version of the directionality problem and will be discussed in detail later (§ 4.6).

[33] Note the connection to the argument for conservation from "existence value" (Chapter 3, §3.2).

[34] The problem here is once again that of directionality: "bad" symbols, those of the past, are supposed to transform human demands in an undesirable way and, therefore, should be destroyed; the "good" new symbols are supposed to transform these demands in a way congruent with the progress of the revolution.

(i) As was briefly indicated in Chapter 2 (§ 2.3), and has been noted by many historians and philosophers before,[35] the appeal of wilderness is of very recent vintage, and largely confined to the North. There are probably multiple sources for this appeal of wilderness. It is possible that it has increased as a result of the increasing urbanization of modern societies. A lack of immediate access to "nature" may affect our psychological states – including our emotional states – in ways that we do not yet scientifically understand. Or it may the case that this appeal reflects the human evolutionary past.[36] Whatever may be the etiology of the appeal of wilderness, for those who have that love, the experience of wilderness is presumably transformative, whether because it is beautiful or because it is sublime. The psychological solace that wilderness provides to many of us is presumably a result of this transformative capacity. That is why we value wilderness and, faced with increasing urbanization and sprawl, argue that the remaining fragments of wilderness should have high priority for preservation. Importantly, to the extent that this argument for wilderness preservation based on transformative value is correct, it remains so even in the light of two otherwise compelling arguments against such preservation. (a) It is argued that nonequilibrial ecological change is ubiquitous and equilibria rare. The so-called ecological argument against wilderness[37] consists of pointing out that preserving wilderness in some definite pristine state requires interfering with ongoing natural processes. At least in their rhetoric, wilderness preservationists claim to be advocating the exact opposite. However, if wildernesses acquire their transformative value because of some deep psychological needs that we have, it is irrelevant whether we have to transform ongoing ecological processes in order to satisfy those needs, that is, to preserve wilderness. There is an interesting irony here: a resolute anthropocentrism justifies interference with nature in defense of wildernesses. (b) Because what is often

[35] See, for example, Nash (1973, 1982). Sarkar (1999) provides a review.

[36] Evolutionary psychologists sometimes trace this appreciation to the environment in which human minds initially evolved (see, for instance, Orians and Heerwagen 1992). Since these were presumably spaces with plenty of scope for solitude, we are supposed to have a deep preference for them "hard-wired" into our brains. Since primate mental evolution did not leave a fossil record, it is impossible to judge the plausibility of such speculative stories. As noted in the text, none of the discussion there hinges on the etiology of the transformative capacities of wildernesses and other natural habitats. This discussion is not intended as an endorsement of evolutionary psychology.

[37] Woods (2001), p. 353, is responsible for the terms "ecological argument" and "no-wilderness argument" in this context.

designated as wilderness comprises landscapes transformed by human presence and activities (recall the discussion of Chapter 2, § 2.3), the so-called no-wilderness argument holds that the absence of genuine wildernesses makes the program of preserving them logically impossible. Once again, if wildernesses are to be preserved because of their transformative value, this objection becomes irrelevant. If some landscape that we have designated as a wilderness (for whatever reason) has the transformative capacity under discussion, it deserves protection because it possesses that capacity, irrespective of its etiology and authenticity as a wilderness. (There may, however, be a problem here: it is possible that the transformative capacity, as in many aesthetic contexts, relies on the presumed authenticity of the entity. In that case, the no-wilderness objection remains serious.)

There is yet another interesting irony in this situation. In Chapter 2 (§ 2.3), wilderness preservation was contrasted with biodiversity conservation. It was pointed out that their divergent goals must be kept distinct, and that there are situations in which the two goals may be in conflict. Yet, from the point of view being developed here, wilderness acquires its value precisely from the same source as biodiversity, from its ability to transform the felt preferences or demand values of individuals. It follows that biodiversity conservationists and wilderness preservationists must respect each others' values and understand that, when there is a conflict, important values are at stake for both sides. At the very most, biodiversity conservationists can argue that wilderness does not offer a counterpart to the intellectual contribution that biodiversity potentially makes, a feature of biodiversity that is an important part of what gives it transformative value. Thus the second method by which biodiversity may acquire transformative value *may* make it more important than wilderness preservation. However, even this is a matter of degree, not a qualitative difference.

(ii) It is arguable – though there are obviously many other aesthetic theories that would admit no such claim – that aesthetic objects also acquire their value because of the power that they have to transform our demand values. This may be because of (a) cultural and other associations that an aesthetic object generates in an individual, (b) the semantic meaning of narratives in some forms of art, (c) less cognitively mediated associations, or even (d) purely affective responses. According to this argument, an aesthetic experience transforms other aspects of our life in a way similar to that produced by some experiences of wilderness and perhaps even by religious or spiritual experiences. From this perspective, for good or for bad, the sharp contrast that is often made between objects fabricated by humans and natural objects disappears. Moreover, for the same reason that we have an obligation to conserve biodiversity or even to preserve wilderness, we have an obligation to preserve works of art. The owner of a Picasso painting may not harm it for the same reason that the owner of a piece of land may not harm an endangered species on it.[38] Recourse to transformative values provides stronger obligations to works of art than most aesthetic theories prevalent today. This may be another virtue of the framework based on transformative value developed here. Nevertheless, it should not go unnoticed (a) that it is at least a weakly reductionist account, ultimately deriving aesthetic values from human demand values, and (b) that it is also a rudimentary naturalized account, grounding the correctness of claims of aesthetic taste on empirical facts of human psychology. (Further discussion of aesthetic issues along these lines, no matter how interesting, would take us beyond the scope of this book – see, however, Chapter 8, § 8.1, where this problem is one of those that are identified as meriting further exploration.)

Returning to the general discussion of biodiversity conservation, it is worth considering whether a justification for such conservation should be constructed on aesthetic grounds rather than on the ethical basis being discussed in this chapter, or at least in addition to it, thus promoting a pluralist defense of conservation.[39] If what was just said about transformative value being the source of aesthetic value is correct, an aesthetic justification for biodiversity would be just as consistent with basing the value of biodiversity on its transformative capacity as our ethical one. Recourse to transformative value also partially, though only partially (see point [i] below), sidesteps the most commonly recognized problem with aesthetic defenses of preservationism: how natural, as opposed to cultural, objects acquire

[38] In the United States, under the provisions of the Endangered Species Act, to do no such harm becomes a legal obligation. What is at stake in this analogy is the basis for such legislation and not merely the legislation itself.

[39] Sober (1986) argues that an aesthetic defense is necessary because of the directionality problem – recall the discussion of § 4.1; the directionality problem will be further discussed in §§ 4.6–4.7. Sagoff (1991) argues for pluralism; Thompson (1995) is among those who think that an aesthetic justification is sufficient. However, neither of these papers adequately distinguishes between nature and biodiversity.

aesthetic value.[40] Moreover, aesthetic considerations have sometimes carried practical weight in discussions of the preservation of nature.

Nevertheless, there are two problems with any attempt to defend biodiversity conservation on aesthetic grounds:

(i) To date, almost all defenses of the aesthetics of preserving nature do not clearly distinguish between nature as a general category and biodiversity. However plausible it may be to ascribe aesthetic properties, such as that of being beautiful or sublime, to certain natural landscapes or features, it is not quite so straightforward to attribute such properties to features of biodiversity. (It is being assumed here that standard aesthetic theories and conceptions of art provide some minimal account of what counts as aesthetic properties.) What, for instance, is the property of the cryptic genetic (allelic) diversity of a population that makes it *aesthetically* valuable? Simply noting that it has transformative value is not a sufficient answer; having transformative value does not, by itself, determine whether an entity should be regarded as having ethical or aesthetic value. Is it plausible that every feature of biodiversity has some property that can be relevantly connected to aesthetics so as to count as an aesthetic property? These are the questions that must be answered before an aesthetic defense of biodiversity conservation can be carried much further. Note, however, that from a pluralist perspective in which aesthetic value is used along with ethical value as a justification for biodiversity conservation, this problem does not require a solution in every case. However, all the hard work, when following such a strategy, is ultimately done by ethical value.

(ii) The aesthetic value of biodiversity must be such that it trumps the utility (in the form of demand value or the satisfaction of felt preferences) that would be realized by its nonconservation (that is, the forgone opportunities).[41] This problem is also faced by ethical defenses of biodiversity, but it is more likely solvable there, at least in the context of public policy formulation, since we usually give higher priority to ethical than aesthetic considerations in such contexts. It is much more difficult to argue that the transformative value of some component of biodiversity, where this value is acquired because of aesthetic properties, suffices to trump ordinary demand value.

[40] For a more extended discussion of this problem, see Fisher (2001) and the references therein.

[41] Sober (1986) is among those who argue that this aesthetic value will probably not routinely trump other utilities. The aesthetic argument for biodiversity conservation thus remains weak, a conclusion that Sober accepts.

4.5. THE BOUNDARY PROBLEM

A serious problem in giving an adequate account of transformative values is not that there are entities other than biodiversity that can have transformative value, but rather that there seems to be no principled way to delimit what can have such value. The examples discussed in Section 4.3 showed that the set of entities that can bear transformative value is highly heterogeneous. There is obviously not much in common between many imperatives: to conserve biodiversity, to have sexual intimacy, to worship at the altar. This is the "boundary problem," that of establishing the boundary of the set of entities that may have transformative value. Consider drug experiences. Few would doubt that these have transformative value. For cultures that have evolved ritual drug-use practices, including the use of those hallucinogenic drugs that apparently terrify governments of the North, the transformative capacity of drugs is used for identifiable social ends, including, but not limited to, the achievement of social cohesion. It is easy to condemn such ritual practices, as is the wont of those who have confidence in the superiority of the one global form of culture that has arisen with late capitalism. However, as cultural anthropologists have shown us for the last half-century, it is hard to find rationally defensible arguments for such blanket condemnation.[42] More often than not, such condemnation reflects nothing more than an assertion of cultural prejudice.

However, we need not invoke anything as controversial as the use of drugs to face the boundary problem. There seems to be no principled way to deny that the most trivial of episodes may have transformative value for some individual. Think of what is, to you, the most trivial genre of music. Now return to the example from Section 4.2 of someone providing an individual with a ticket to a concert, and assume that it is a concert featuring music of that genre. The argument that was given in that section shows that even this concert potentially has transformative value. It is possible that an individual might attend such a concert and return with transformed demand values for that, and other, genres of music. Nothing that has been said so far argues against such an eventuality.

A certain type of mystic revels in the transformative value of a blade of ordinary grass. We will return to this blade of grass several times later. But it does not take a mystic to broach such possibilities. Existential philosophers – though not all philosophers in the phenomenological tradition – have often reveled in such cases. In Sartre's *Nausea*, Antoine Roquentin is transfixed

[42] See, for example, Siskind (1973) and Schultes (1976).

by an experience of the "being" of some roots.[43] This is a fictional example, but there is no reason, in principle, to believe that such experiences cannot occur. Can we exclude these kinds of frivolous entities as bearers of transformative values? It does not appear to be easy. Must we do so? We must if we want to avoid being forced into a position where we have to defend the preservation of almost anything on the ground that it may bear transformative value for some individual somewhere. Such a position may have appeal for those who, in general, advocate a certain quietism toward life, but in the context of biodiversity conservation it is unacceptable, for two reasons: (i) an adequate conservationist ethic must establish some value for biodiversity over some arbitrary candidate for preservation; (ii) against quietism, an adequate conservationist ethic must provide a rationale for – and even demand – intervention. We cannot avoid the boundary problem. Moreover, an appeal to some number of individuals who must be subject to transformation by an experience – for instance, by requiring that a majority of the population be so transformed – is of little help. We do not know how much of the population is likely to have its demand values transformed by an experience of biodiversity. We do not even know if this fraction is greater than that which may benefit economically from the destruction of biodiversity. And we have no nonarbitrary way to choose the number in question.

4.6. THE DIRECTIONALITY PROBLEM[44]

To make matters worse, we have yet to solve the directionality problem, which now arises again, but in a manner slightly different from the way it arose at the beginning of this chapter: transformative experiences may be negative, just as they may be positive. Here, by a "positive" effect what is meant is that the results are desirable on ethical grounds, and not necessarily that the individual has an increased desire (or felt preference) for that experience. An experience of child pornography may increase an individual's desire for it; however, this is a negative effect. Return to the example

[43] See Sartre (1949). Or, for that matter, see Blake's invitation in "Auguries of Innocence":

> To see the World in a Grain of Sand
> And Heaven in a Wild Flower,
> Hold Infinity in the Palm of Your hand
> And Eternity in an hour."

See <http://www.cs.rice.edu/~ssiyer/minstrels/poems/368.html>.

[44] See also Norton (1987), p. 11, where this problem is briefly discussed but does not get the attention it deserves.

of drug use. Few would deny that drug experiences have the potential to be immensely negative and to do serious, potentially permanent, psychological harm to an individual. This, at least, is one valid source of the worry that contemporary Northern societies have about drug use. Ritual use of drugs in societies that allow such use typically involves a degree of social guidance and, more importantly, social support that has presumably evolved culturally to ensure that these ritual drug experiences are seldom negative. There is anecdotal evidence to suggest that even when drugs were being routinely experimented with in Northern societies – a generation ago – groups evolved practices that provided support to initiates. Recreational drugs were usually used in groups, not alone.[45]

Even experience of biodiversity can have potentially negative transformative value. Imagine an individual visiting a wildlife preserve for the first time. An accidental encounter with a dangerous wild animal, such as a venomous snake, can potentially be terrifying. It may even lead to a lack of sympathy for that animal's conservation. What is important in this context is that an experience of this sort may result in a change in the felt preferences of an individual in a way that affects biodiversity conservation negatively even more generally. It may decrease the probability of that individual's contributing to any kind of biodiversity conservation because of a new dislike of dangerous animals in the wild. In such a situation, an encounter with biodiversity will have had a negative transformative value with respect to biodiversity conservation. (Wildernesses and religious practices may generate negative transformative value through unpleasant experiences in exactly the same way.)

Even experiences of biodiversity that we normally take to be unproblematically positive may have a potential for generating negative transformative value in a way that is inimical to biodiversity conservation. Neotropical rainforests are notoriously safe for humans compared, for instance, to rainforests elsewhere in the world. There are no mammalian predators of humans, few dangerous snakes or insects. Yet those who tend to be claustrophobic may well find the interior of a neotropical rainforest distasteful, if not worse. They may react negatively to the absence of bright light, to the constant humidity, and so on. They may return from a visit to a rainforest wondering what its appeal is. Different individuals have different tastes.

In general, sublime experiences can easily be terrifying in a way that we may regard as harmful. Whether it be in the form of biodiversity, wilderness,

45 For a discussion, see Maisto (2003), though that discussion is largely limited to the context of the United States.

or drugs, an experience of the sublime carries with it a certain psychological risk for the individual. In a sense, this is part of what makes an experience sublime. The best that we can say is that, behaviorally, the pursuit of the sublime through dangerous thrill-seeking activities continues to be a part of human culture. While this may be a sociologically interesting generalization, it does not solve the *normative* directionality problem.

4.7. SOLUTIONS

Are there solutions to the boundary and directionality problems? If by a solution what is meant is a sound argument that shows that these problems are not genuine problems but merely arise from mistaken reasoning, there is no solution. Nevertheless, there are reasons to suggest that the problems can be mitigated to the extent that they do not vitiate a defense of biodiversity conservation on the basis of the transformative value of biodiversity. Beginning with the latter problem, note that we do not have to solve the directionality problem for all entities that have transformative value. We merely have to solve it for biodiversity and establish that it is much more likely to have positive than negative transformative value. It is irrelevant whether the particular argument we present carries over to other entities such as wilderness, religion, or drugs – as a matter of fact, it does not.

Recall the two ways in which biodiversity acquired its transformative value: (i) directly, through individuals' encounters with biodiversity, and (ii) indirectly, because of its potential intellectual contributions. Now recall, reflecting on what was said in the last section, that it is usually the case – one exception will be discussed in the next paragraph – that biodiversity can potentially have negative transformative value only through individuals' unfortunate direct encounters with it. But how likely are such encounters? Going by what empirical evidence we have, not very. For every negatively transformative human encounter with any aspect of biodiversity, there are probably thousands that are positive.[46] (One can probably make a similar case for wilderness, a less plausible one for religion, probably no case at all for drugs, given the patterns of their consumption that have developed in Northern societies as opposed to societies with a long tradition of drug use.) This argument can also potentially be extended to the Madagascar periwinkle versus the Ebola virus case discussed earlier in this chapter, though perhaps not quite so easily. There are far too many biotic pathogens.

[46] See Kellert (1996), Chapter 7, for a discussion geared to endangered species in the United States.

(If it can be so extended, then the case for biodiversity conservation purely on the basis of demand values would be strengthened. However, the problem of forgone opportunities – or the cost of conservation – would still have to be solved.)

The most convincing argument in response to the directionality problem is based on the indirect transformation of demand values that biodiversity potentially generates through its contributions to science.[47] The only potentially negative contribution that biodiversity can make in this context seems to be through the misuses of science that it may enable, for instance, through novel possibilities for biological warfare. There is little doubt that the use of biodiversity studies to discover new pathogens and produce new biological weapons – an unfortunately tangible possibility – would lead to transformations of preferences that are undesirable under any conceivable ethical scheme. However, biodiversity is hardly unique in facing such a predicament. Anything that contributes to science does. The creation of quantum mechanics and relativity theory, two of the twentieth century's greatest intellectual achievements, made possible the nuclear bombs that were dropped on Hiroshima and Nagasaki.[48] In fact, biodiversity studies – for instance, studies of natural history – suffer far less from this problem than almost any other scientific discipline. Even the creation of new biological pathogens requires more help from molecular biology than from biodiversity studies. One may choose to conclude from the potential for the political misuse of science that all scientific endeavor suffers from a directionality problem with no resolution. A far more reasonable conclusion is that science requires sociopolitical oversight, even though the good done by science still far outweighs the harm. Meanwhile, given all the positive contributions biodiversity has made to science, the transformative value of biodiversity must on the whole be overwhelmingly positive.

The argument just given is not fully satisfactory on two grounds:

(i) It does not offer a general account of what sorts of transformations are positive. Rather, it relies on the assumption that scientific knowledge as a whole has positive transformative value and exploits the connection between biodiversity and science as a way to co-opt that value for biodiversity. For those who accept the claim that the transformative value of science is positive, this move will be sufficient. But for those who do

[47] Only a very few of these were noted in § 4.2. Recall also the discussion of the myth of lost futures in Chapter 2 (§ 2.1), where many other examples were given.

[48] For the sake of argument, it is being assumed – though there are also compelling contrary considerations – that dropping these bombs was ethically undesirable.

not accept this claim without question, this move is question-begging. They will insist on a general account of positive transformative value. Recall that transformative values transform demand values and that the latter reflect felt preferences. Norton distinguishes between such felt preferences and "considered" preferences, which are "hypothetical desires or needs. They are desires or needs an individual *would have* if certain very stringent, perhaps even practically impossible, conditions were fulfilled."[49] A considered preference is the result of careful reflection, which is "taken to include a judgment that the desire or need is consistent with a rationally adopted worldview, which in turn includes a set of fully defended scientific theories, and a set of rationally developed, fully defended aesthetic and moral ideals."[50] If the intellectual interest of biodiversity (and of science in general) gives it its transformative value, then clearly it satisfies the requirements of Norton's account, provided that, bearing in mind the fallibility of any scientific theory or model, we interpret "fully defended" rather liberally. Otherwise, Norton's account is too stringent, and the judgment that a particular experience has positive transformative value will not be epistemologically accessible to us. The difficulty is to set adequate constraints on the liberality with which Norton's criterion should be interpreted in practice. Yet another problem with this account is that it refers to "aesthetic and moral ideals," which now require some sort of justification that is not circular, that is, does not assume that the "rational" development of these ideals does not automatically include the conservation of biodiversity.

An alternative solution to the directionality problem, and one that is more promising – though not fully satisfactory – is to attempt to lay down precise conditions for an experience to have positive, rather than negative, transformative value. There are two of these that are probably *jointly* sufficient:[51] (a) the experience is self-enhancing in the sense that having the experience generates a felt preference for the experience;[52] and (b) there is no overall actual and potential harm associated with the experience (with harm defined using felt preferences). An accidental

[49] See Norton (1987), p. 9; emphasis in the original.

[50] Ibid.

[51] In this discussion, it is being assumed that it has already been established that the experience in question has transformative value; all that is yet to be decided is whether that value is positive or negative.

[52] Such self-enhancement is a hallmark of good scientific research: successful science encourages further scientific exploration.

encounter with a venomous snake violates both (a) and (b). Experiencing child pornography probably satisfies (a) for some individuals but falls afoul of (b). (For many, adult pornography has the same status, but the matter is much more controversial.) Scientific research, as well as biodiversity, satisfy both. Notice that condition (b) is weak: it does not require overall good; it only requires no harm. Biodiversity probably satisfies condition (b) less controversially than all scientific research, given that some forms of scientific research have been geared to ends such as weapons production. However, biodiversity sometimes, and wilderness often, falls afoul of (b) if preservation involves the denial of resources to human inhabitants and users of places – this is the sense in which this solution is yet not fully satisfactory. These two criteria by no means provide a complete account of positive transformative value. However, they do provide a beginning that suffices for the problem at hand.

(ii) By claiming such a close relationship between the pursuit of science and the conservation of biodiversity, the argument just given, in a sense, makes the ultimate rationale for biodiversity conservation its scientific value. For many, including some deep ecologists and others who believe in the intrinsic value of biodiversity, its conservation is invaluable irrespective of its use to science.[53] Some of the difficulties with this position have already been noted in the last chapter. Moreover, those who make such a claim are usually conflating biodiversity with nature and are not treating the former in the technical sense that is being used here – see also Chapter 6 (§ 6.5).

Turning to the boundary problem, one potential response that is available is to say that, now that the directionality problem is assuaged, the boundary problem is irrelevant. We have established that, by and large, biodiversity has a positive transformative value. Since all that we had to establish is this case for biodiversity, we may choose to say that it simply does not matter if other entities also have positive transformative value. But this is not enough. We have to establish that biodiversity has transformative value in some way that the mystic's blade of grass does not, or at the very least, that its positive transformative value is significantly greater than that of the mystic's blade of grass. Otherwise, conserving the mystic's blade of grass may well deserve higher social priority than conserving what remains of, say, the

[53] See, for instance, Naess (1986). Recall the discussion of deep ecology in Chapter 3 (§ 3.4).

almost-vanished South Asian tropical rainforests. This is a nontrivial problem but, once again, not one that is insurmountable.

As noted in the last section, we have no principled way of denying that the mystic's blade of grass may have transformative value, and we will not follow the route of attempting to do so. Rather, the argument will be that such entities have only an "incidental" transformative value that is irrelevant in the context of policy formulation when compared to the "systematic" transformative value of entities such as biodiversity (or, for that matter, wilderness). An entity has only "incidental" transformative value if it satisfies either of two related criteria:

(i) if some other entity can be substituted for it without loss of that value, to the extent that we can determine this by some systematic procedure, or
(ii) if only some small set of entities provide the requisite transformations, and if we have no reason to believe that this set can be indefinitely enlarged following some systematic procedure.

The mystic's blades of grass satisfy both criteria. It may well be the case that a grain of sand, or another blade of grass, will transform the mystic who was transformed in the first instance. Yet we have no reason to believe that the next potential mystic will be transformed by the next blade of grass (or anything at all) – this, after all, is why *mystics* alone have such experiences. What the blade of grass achieves has more to do with the mystic than with the blade itself.

An entity has "systematic" transformative value if we have reasons for giving it this value other than the mere fact that some individual stood transformed by it: we have a generalizable account of how it acquired such a value. From this account of the acquisition of transformative value, we will generally be able to determine when an entity can be substituted for another. Ideally, our account should also tell us how to add entities indefinitely to the set of those that already have transformative value. (That is the point of requiring that the account be generalizable.) We can thus build a larger set from an initial one.

For biodiversity, such an account was given earlier in the chapter when the two ways in which it may transform our demand values were explicated. Referring to the second and more important way, we know that a biological entity has transformative value when it has the potential to contribute to science. In that context, for instance, most individuals of a parthenogenetic species may be substituted for others that are genetically identical. However, the species as a whole may not be so substituted. We also know how to add to the set of entities with transformative value by judging whether a new entity will potentially contribute something new to science, for instance, whether it is a novel or unusual form of life. Every scientific novelty has such a potential to some extent. Given how much we have yet to learn about the variety of life on Earth, biodiversity studies have more potential in this way than probably any other field. Thus, biodiversity satisfies our definition and has systematic transformative value. Moreover, simply because novel phenomena are always more likely to contribute to science than those that are not, this argument also provides a rationale for our usual assumption that new or rare biodiversity components deserve more attention than common ones. The solution to the boundary problem is simply the following: when formulating policies of preservation, that is, when selecting entities for conservation action, include everything that has systematic transformative value and exclude anything that does not, including those entities that have incidental transformative value.

4.8. ADEQUACY TESTS

The conservationist ethic developed in this chapter trivially satisfies the first four adequacy conditions (generality, moral force, collectivity, and all-taxa) listed near the beginning of Chapter 3 (§ 3.1). The generality condition ([i] of § 3.1) required that an adequate conservationist ethic give value to biodiversity in general; our ethic does exactly that. The moral force condition ([ii] of § 3.1) required that the ethic produce an obligation to conserve biodiversity – the arguments of Section 4.3 were designed to show how this account does so. The collectivity condition ([iii] of § 3.1) required that the ethic give value at least to species and populations; the all-taxa condition ([iv] of § 3.1) required that it give value to all species, not just to charismatic or otherwise selected ones. Our ethic satisfies both these conditions.[54] Now, the non-anthropocentrism condition ([vi] of § 3.1) demanded that, preferably, the ethic should not be based on parochial human values. Whether or not this condition is also satisfied depends on whether the demand values that are transformed by biodiversity should be regarded as "parochial."

[54] It can plausibly be argued that charismatic species have more transformative value than uncharismatic ones. However, this is only true if we assume that the main way to acquire transformative value is directly, through those transformations that result from encounters with biodiversity. As was emphasized earlier in the chapter, the main way in which biodiversity acquires transformative value is indirectly, through its capacity to transform our scientific worldview. This does not favor charismatic species. In fact, it militates against them, since they are most likely to be rather well-studied.

(Obviously they are human values.) Not much depends on how we answer this question, since the non-anthropocentrism condition was not given a status equal to that of the others. The non-anthropocentrism condition was introduced only to capture the reverence for biodiversity that most conservationists have and that has led many of them to attribute intrinsic (intrinsic$_1$ or intrinsic$_2$) value to biodiversity. If attribution of transformative value already captures that reverence, the non-anthropocentrism condition becomes irrelevant. To claim that biodiversity has the type of transformative power that was described earlier in this chapter is to show considerable reverence for biodiversity.

The case of the priority-setting condition ([v] of § 3.1) is more interesting. At the very least, according to the priority-setting condition, an adequate conservationist ethic must permit the prioritization of species (or other surrogates for biodiversity) for conservation. Preferably, it must provide a method for such prioritization. Moreover, it should validate the conservation actions we typically want, for instance, the prioritization of conservation action directed at endangered and threatened species over action directed at those that are not, or our paying more attention to a species that is unique in its genus than to a species with many congenerics.

As indicated at the end of the last section, the conservationist ethic in this chapter does precisely that. Endangered species have a priority over those that are not because the former are more likely to become extinct than the latter. Every species has transformative value because of both the potential experiences they provide for us and their scientific interest. However, if there is a finite amount of resources available for conservation, and if these resources are expended equally on all species, the endangered ones may still be at an unacceptable risk of extinction, whereas it is unlikely that there will be a significant change in the status (with respect to extinction) of nonendangered species. However, if the resources are expended preferentially on endangered species, there is often a high probability that they can be brought back from the brink of extinction. Meanwhile, little, if any, harm would probably have been done to the nonendangered species. Consequently, an attempt to conserve all species involves a stronger obligation to direct conservation actions toward endangered species than toward nonendangered ones. Further, the same argument shows that this ethic also justifies conserving species that are biologically different over those that are similar. These are more likely to contribute to future scientific developments. Thus it will justify paying attention to diversity at taxonomic levels higher than that of species.

Moreover, species that we do not know, living in habitats that have not been fully explored, receive very high priority. These are entities the experience of which may well be of a sort that is entirely new to us. Moreover, they are most likely to contribute novel insights into the biological world. In 1995, for instance, a new species, *Symbion pandora*, belonging to an entirely new phylum, Cycliophora, was discovered in the bristles of the well-known Norway lobster (*Nephrops norvegicus*).[55] From a scientific point of view, the discovery is particularly important. We do not yet have a good understanding of the rules of morphological development, especially of what restricts the body plans of plants and animals. We know, for instance, that all adult animals have either bilateral or fivefold symmetry. But we do not know why. We do not know why other types of symmetry are never seen, whether there are developmental or evolutionary rules preventing the formation of such organisms.[56] Each phylum represents a different body plan. Every new phylum helps assemble part of the puzzle.

To the extent that the progress of science transforms demand values, the discovery of the new Cycliophora phylum must thus be regarded as a significant contribution in that direction. An organism that is so novel in structure and life cycle as to require a new phylum of its own had gone unnoticed in spite of being very easily accessible to biologists. It is hardly surprising, then, that unexplored regions of the world (including the oceans) are believed to harbor many surprises. Given their high potential to transform our knowledge and, directly and indirectly, our demand values, these are the regions that receive the highest priority according to the ethic developed in this chapter. Every effort should be expended to conserve them. Recalling the discussion of Chapter 2, there is something potentially quite valuable that is captured by the myth of lost futures (§ 2.1). If new phyla can be discovered in 1995, there is no justifiable fear of diminishing returns from our study – and enjoyment – of the diversity of life around us. We finally have a conservation ethic that satisfies our initial conditions for adequacy, even though many problems, such as those discussed earlier, remain to be sorted out.

[55] See Funch and Kristensen (1995) for the report; Morris (1995) discusses the significance of this discovery.

[56] For a philosophical perspective on the growing literature tying development and evolution, see Robert (2004).

5

Problems of Ecology

As the ecologists Yrjö Haila and Richard Levins have pointed out, "ecology" has at least four meanings, each of which is relevant in the context of discussions of contemporary biodiversity conservation and other environmental problems:

- ecology *the nature*: nature's economy as a material fact, and as a material basis for human existence;
- ecology *the science*: the biological discipline investigating nature's economy;
- ecology *the idea*: prescriptive views of human existence, derived from what is known or believed about nature's economy; and
- ecology *the movement*: political activities trying to transform society to agree with ecological ideals.[1]

If the first use of "ecology" is meant to imply that substantive claims can be made about the "material" facts of ecology, independent of the scientific or descriptive framework invoked in the second use of "ecology" as a science, the distinction made between these two uses involves an acceptance of at least a modest version of the philosophical doctrine of *realism*: the claim that it is possible to make claims about the world independent of our descriptions of it. In the interest of avoiding contentious philosophical issues that do not directly impinge on the aims of this book, no position will be taken here on the tenability of this doctrine. Rather, except when otherwise indicated, "ecology" will be used in the second sense. Moreover, because it will be assumed that biologically correct descriptions are those that withstand empirical testing against nature,

[1] Haila and Levins (1992), p. ix (emphasis in the original).

whatever value the first use has will be implicitly subsumed under the second.[2]

To say that there are prescriptive views about human existence that can be derived from our biological knowledge of ecology comes dangerously close to embracing yet another highly controversial philosophical position, which many philosophers consider to be based on a fallacy. This supposed fallacy lies in the claim that a normative injunction (a prescription, that is, an "ought" statement) can be derived from a descriptive one (an "is" statement).[3] Haila and Levin do not intend any such ambitious derivation; they explicitly reject the idea that scientific ecology forces any normative conclusions on us. Rather, the prescriptions they have in mind are injunctions that remind us, as we devise strategies for our survival and well-being that often have the form of prescriptive ought-statements, that we are constrained by ecological facts. For instance, all living organisms require adequate nutrition; nutrition affects not only the ability of individual organisms to continue living, but also their reproductive capabilities; human well-being almost certainly requires a rich cultural and social life, and so on. We can easily accept such injunctions without taking any position on whether the supposed fallacy just mentioned is, indeed, a fallacy.[4]

Rather, acceptance of such injunctions involves reasoning of the following sort. Descriptive ecology may give us a purely descriptive statement such as:

(i) If wetland *A* is drained, species *X* will become extinct.

Now suppose that we believe:

(ii) It would be wrong to drive species *X* to extinction.

From (ii), along with (i), we would conclude:

(iii) It would be wrong to drain wetland *A*.

Claims (ii) and (iii) are both normative. In this argument, claim (iii) is being derived from claims (i) and (ii). However, because claim (ii) is normative, we

[2] In philosophical terms, what will be assumed is that a scientific model should be empirically adequate; this assumption involves no commitment to realism.

[3] This is sometimes called the "naturalistic fallacy" (see, for instance, Horgan 1995, p. 176) though most philosophers usually reserve that term for a different claim, that ethical properties cannot be defined on the basis of nonethical properties (Moore 1903; see Frankena 1939 for a careful discussion of the differences and relations between the two fallacies).

[4] This move also reflects the earlier decision not to take potentially controversial philosophical stances that are not forced upon us by the contexts of this book.

are not deriving a normative claim from purely descriptive ones. Whatever normativity claim (iii) has is derived from the normativity already present in claim (ii).[5] It will not be assumed here that these prescriptive injunctions also form a doctrine that can appropriately be called "ecology"; the term will not be used in Haila and Levins's third sense in this book. In fact, this third use of "ecology" is relatively rare in any context.

It deserves emphasis that nothing peculiar is going on in the argument given in the last paragraph. The use of descriptive facts – along with normative assumptions – is typical of normative, particularly moral, reasoning. To say that we should not aim a bomb at a maternity ward because we think that it is wrong to kill innocent babies assumes the descriptive fact that bursting bombs tend to kill innocent babies in their vicinity. In most moral contexts, relatively trivial – and uncontroversial – descriptive facts constitute the descriptive repertoire we use in our normative reasoning. When the descriptive facts become uncertain – as there is uncertainty, for instance, about the appropriate definition of the beginning of human life in the context of debates over abortion – moral disagreements and controversies emerge. What is rather unique about environmental contexts is that the descriptive repertoire includes a rather sophisticated body of knowledge that is recognized as a science by itself: the traditional science of ecology and parts of the new discipline of conservation biology.[6] The major thrust of this chapter will be to examine exactly the ways and the extent to which the science of ecology can contribute to our normative considerations when devising environmental policy. Conceptual problems and technical limitations will be emphasized, but not because of any disdain for ecology's achievements and status as a science; rather, this focus is intended to encourage further conceptual and empirical work that should result in better science and more successful environmental planning. This chapter is not intended as a résumé of all philosophical problems associated with the science of ecology. It considers only those problems that are germane to biodiversity conservation.

The fourth use of "ecology" has become commonplace during the last thirty years: we are confronted with "deep ecology," "liberation ecology," "political ecology," and "social ecology," just to mention the most common political ideologies associated in one way or other with environmental

[5] In the terminology of Wood (2000), pp. 4–12, the normative claim (ii) provides the "operative reason," whereas the descriptive claim (ii) – like all other empirical claims in the context of decision making – constitutes only an "auxiliary reason."

[6] The other case that is similar in this respect is medical ethics. Ordinary political ethics does not count, because there is hardly a sophisticated body of knowledge that genuinely counts as a "science" in political "science."

concerns.[7] To some extent, each of these ideologies has inspired political movements based on them. Some of these ideologies have already been mentioned in earlier chapters; others will occasionally be mentioned later in this book. The presence of the appropriate qualifier ("deep," "liberation," etc.) will explicitly indicate that there has been a shift from "ecology" as a science to "ecology" as a political ideology or movement.

5.1. ECOLOGICAL MODELS

How is scientific ecology relevant to environmental concerns? The term "ecology" was coined by the German zoologist Ernst Haeckel in 1866 to describe the "economies" of living forms.[8] The theoretical practice of ecology consists, by and large, of the construction of models of the interaction of living systems with their environment (including other living systems). These models are then tested both in the laboratory and in the field. (Fieldwork may also consist of data collection that is not inspired by any theory.) Theory in ecology consists of the heuristics – or principles – used to construct models. Unlike evolutionary theory, ecology has no generally accepted global principles such as Mendel's (and other) rules of genetic inheritance.[9] While ecology is a rapidly changing field with many new subdisciplines such as landscape, spatial, metapopulation, and metacommunity ecology that are rapidly becoming foci of both theoretical and experimental attention, traditionally, especially in the context of biodiversity conservation, three types of models have been the most important:

(i) *Population models*: these are based on representing an ecological system as the set of populations it consists of. Each population, in turn, consists of potentially interacting individuals of a species. Populations

[7] For "deep ecology," recall the discussion of deep ecology in Chapter 3 (§ 3.4). For "liberation ecology," see Peet and Watts (1996). For "political ecology," see Cockburn and Ridgeway (1979); the term goes back at least to Wolf (1972). "Social ecology" was originally introduced in India in the 1940s to describe a scientific research program that consisted of treating human social and natural environmental systems simultaneously. It has always had an overt left-wing political orientation – see Mukerjee (1942) and Guha (1994). Some earlier uses of the term – for instance, that of Alihan (1938) – identify social ecology with human ecology and have no explicit political orientation. A rather different use of "social ecology" is associated, mainly in the North, with Murray Bookchin's project of dialectical naturalism (see, for example, Bookchin 1995); see also Light (1998).

[8] "Ecology" was but one of many terms that Haeckel introduced – one historian calls him the "busiest name-maker of his time"; see Worster (1994), p. 192. For a history of ecology with an emphasis on population ecology, see Kingsland (1985).

[9] Some philosophers find this assessment of theory in ecology overly pessimistic; see Shrader-Frechette and McCoy (1993) and Sarkar (1996) for elaboration of this point.

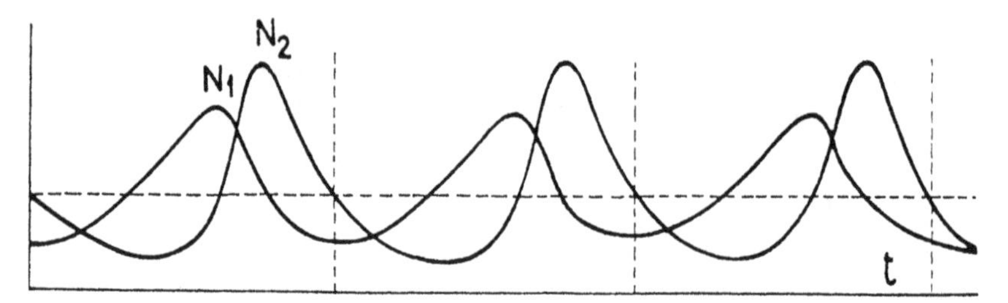

Figure 5.1.1. Predator-prey population cycles. The model, consisting (roughly) of two coupled partial differential equations tracking the size of predator and prey populations, is due to Volterra ([1927] 1978), pp. 80–100; the diagram is from p. 100. There are two species, a predator species with a population N_2, which feeds only on a single prey species with population N_1. The model incorporates demographic stochasticity, which, nevertheless, does not stamp out the basic cyclic pattern. (t is a measure of time.)

may be characterized by their *state* variables (parameters representing properties of the population as a whole – size, density, growth rate, etc.) or by *individual* variables representing the properties of the individuals in them (individual age, fecundity, interactions, etc.).[10] Usually a model considers members of a single or a very few interacting species, for instance, a few predator species and a few prey species. A typical result is that predator-prey interactions lead to population cycles, with the predator population cycle temporally tracking the prey population cycle. The explanation of this phenomenon is straightforward: as prey populations increase, the increased availability of resources allows a rise in predator populations a little later in time. But the increase of predators leads to an increase of prey consumption and, consequently, a decrease in prey populations. But now the lack of resources leads to a decline in predator populations. As predator populations decline, prey populations recover, initiating the cycle once again (see Figure 5.1.1).

From the perspective of biodiversity conservation, the central issue of interest is the changes in the size of populations over time. Population models can be broadly classified into two categories. (a) *Deterministic* models – if population sizes are large, they can be studied using deterministic models, that is, fluctuations in populations sizes due to chance factors (such as accidental deaths) can be ignored. For single species, two standard models are that of exponential and logistic growth (see Box 5.1.1). The former is supposed to capture the behavior of a

[10] Interactions between individuals are also being regarded as properties of individuals; this is permissible so long as the interactions between an individual X and all other individuals in the population can be "averagely" represented as a property of X that does not refer to any other individual (for instance, through a mean-field approximation).

BOX 5.1.1 EXPONENTIAL AND LOGISTIC GROWTH MODELS

Let a population consist of n individuals at time t. Suppose that, in an infinitesimal time interval between t and $t + dt$, a fraction bn of individuals give birth and a fraction dn die. Let the change in the size of the population be dn. Then

$$dn = (b - d)ndt.$$

Let $r = b - d$. Then we get the growth equation:

$$\frac{dn}{dt} = rn.$$

This is the exponential growth model. It assumes that no resource limitation constrains the "intrinsic growth rate," r. It can be solved to give:

$$n(t) = n_0\mathbf{e}^{rt},$$

where n_0 is the size of the population at time $t = 0$.

One way to modify the exponential growth model to incorporate resource limitation is to replace the growth equation of the exponential model by that of the logistic growth model:

$$\frac{dn}{dt} = rn\left(1 - \frac{n}{k}\right),$$

where K is called the "carrying capacity" of the environment; this parameter is supposed to incorporate how resource limitation affects population growth by regulating it. When $n = K$, the growth rate,

$$\frac{dn}{dt} = 0,$$

and the population does not grow any further. Moreover, when there is no resource limitation – that is, $K \rightarrow \infty$ – this model reduces to the exponential growth model. Figure 5.1.2 shows how a population governed by the logistic equation grows in size. At the level of individual behavior, this model does not have the kind of justification that the exponential growth model does. In this sense, it is a purely "phenomenological" model.

population when there is no resource limitation; the latter is one of the simplest ways to try to capture the self-regulation, of population sizes when there is such a limitation. In general, it is assumed that all populations regulate their sizes through mechanisms that are internal

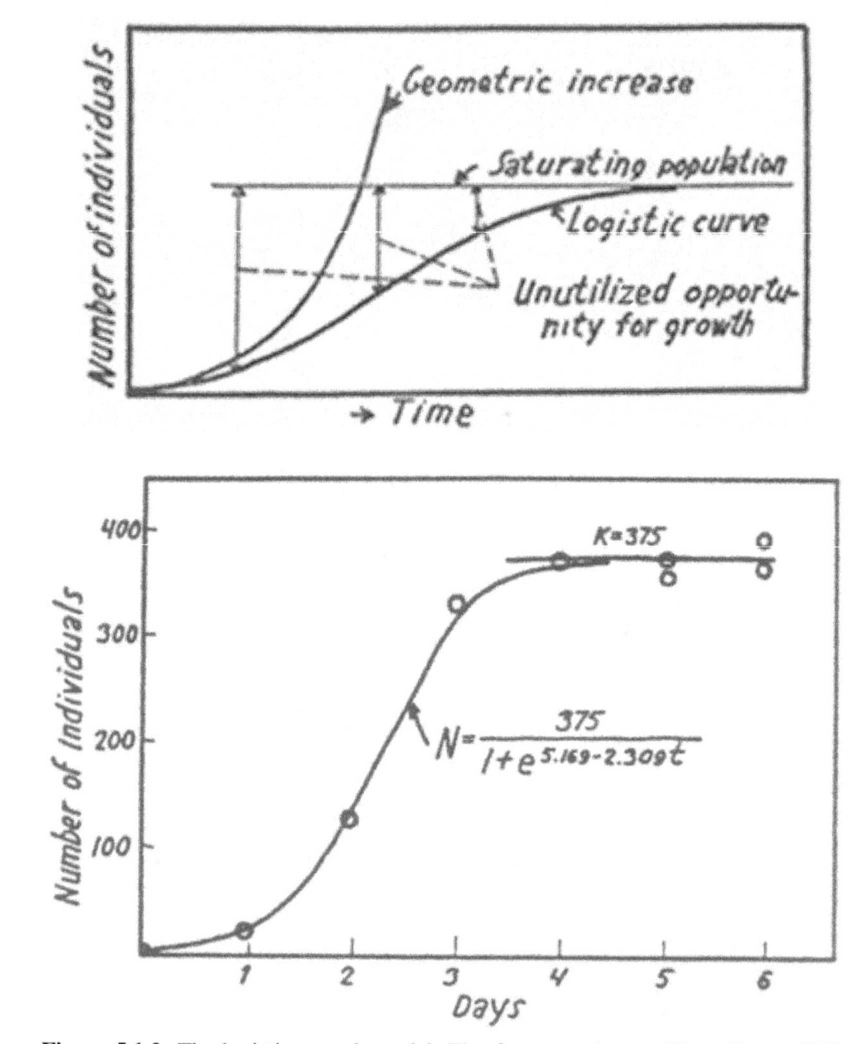

Figure 5.1.2. The logistic growth model. The figure on the top (from Gause 1934, p. 35) shows theoretical curves. "Geometric increase" represents the exponential growth model discussed in the text; "saturating population" refers to the carrying capacity (see Box 5.1.1). The figure on the bottom (from Gause 1934, p. 36) shows an example of an empirical growth curve obtained in the laboratory. If the curve is fitted to a logistic curve (to which it shows similarity), then $K = 375$ is the estimated carrying capacity.

to the population, that is, they show self-regulation. Theoretical exploration of models has made it clear that a wide variety of mechanisms can lead to such self-regulation, though there is as yet no general theory. The precise mechanisms that are playing self-regulative, roles in individual cases are often particularly hard to determine in the field. (b) *Stochastic* models – if population sizes are small, then models must be stochastic: the effects of fluctuations must be explicitly analyzed. The extinction of a population often involves a synergistic interaction between deterministic and stochastic factors. The former lead to the decline of a population to a size that is small enough for stochasticity to become important. Once a population is that small, chance alone may drive it to extinction.

Both deterministic and stochastic models are relevant for biodiversity conservation when they are used to predict the viability of a population, that is, whether it will become extinct within a specified time period. Attempts to assess viability in this fashion constitute population viability analysis (PVA). In the case of deterministic models, an average decline of the size of a population is a harbinger of extinction. In the case of stochastic models, the situation is more complicated: even a population that is, on the average, increasing in size may become extinct because a random fluctuation might reduce its size to zero, from which it can never recover. Usually, three types of stochasticity are distinguished, (a) demographic, (b) environmental, and (c) random catastrophic.[11] The first of these captures chance fluctuations in births and deaths in a population (that is, the so-called vital parameters); the second captures fluctuations in environmental parameters (including those describing the physical habitat, competitors, predators, parasites, etc.); the third captures relatively rare chance events that occur randomly over time, such as floods, fires, and even asteroid impacts. However, short of an explicit mathematical model in which these distinctions are made exact through formal definitions, this classification suffers from ambiguity.[12] Environmental fluctuations and even random catastrophes affect the size of a population only insofar as they affect reproduction

[11] Genetic stochasticity is also sometimes presented as a category in this classification (see Shaffer 1981). However, this practice is conceptually misleading. Genetic stochasticity is a different type of phenomenon from the other three. In small populations, random fluctuations may lead to the random fixation of a gene (allele). Whether this affects the viability of a population depends on what that gene does. Nothing, in principle, even prevents the stochastic fixation of a gene that increases fitness.

[12] Even with a mathematical model, these definitions (like all definitions) are in part conventional and are acceptable only to the extent that they are useful. Usually, models of demographic

and death rates, that is, by creating demographic fluctuations. At least in this sense, the second and third categories of stochasticity are not conceptually independent of the first. Moreover, random catastrophes can well be regarded as extreme cases of environmental stochasticity.

The mathematical analysis of these models is nontrivial. The stated goal of PVA used to be an attempt to estimate the size of the minimum viable population (MVP) that has a specified probability of persisting for a specified number of years. In recent years, however, PVA has moved away from attempts to determine MVPs; this development will be further discussed in Chapter 6 (§ 6.4). The most general and relatively uncontroversial theoretical result to date is that progressively larger populations are required for safety in the face of demographic, environmental, and random catastrophic stochasticity.

(ii) *Community models*: these are models of interacting species, forming an ecological "community," in which each species is treated as a unit. The appropriate definition of "community" has been widely debated among ecologists and philosophers; what is being given here is an interactive definition.[13] Alternative options include defining community by mere geographical association of species at one extreme, or by requiring a high degree of structure in the interactions at the other, making the community analogous to an organism. The simple interactive definition given here is appropriate for two reasons: (a) mere association leaves little of theoretical or practical interest to study or model, while requiring some specified elevated level of interaction would introduce an unnecessary arbitrariness in to the definition; and (b) the former would make any association of species a community, whereas the latter would typically introduce so much structure that virtually no association would constitute a community.

Community models can typically be conveniently represented as loop diagrams,[14] with each species as a vertex and edges connecting these vertices where the species interact. The edges indicate whether the relevant species benefit or are harmed by the interaction, that is, whether they tend to increase or decrease in abundance as a resut

stochasticity are distinguished from models of environmental stochasticity using as a criterion whether the stochastic factor explicitly depends on the population size. If it does, the model in question is one of demographic stochasticity; if it does not, it is one of environmental stochasticity.

[13] See Odenbaugh (2005) for a systematic discussion of the alternatives.

[14] See Levins (1974, 1975), where loop analysis is first introduced; Puccia and Levins (1985) further develop the theory.

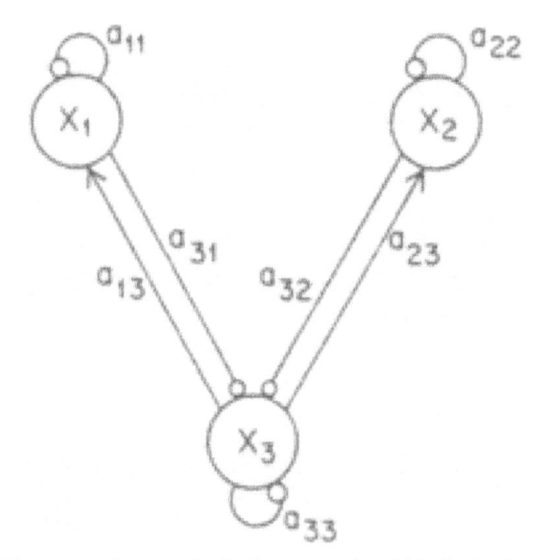

Figure 5.1.3. Structure of an ecological community. This is the loop diagram of a community of three species (from Diamond 1975a, p. 435). Species X_1 and X_2 both prey upon the resource species X_3. There is resource coupling and competitive exclusion. It is assumed that all three species are self-regulating. → indicates positive interaction; –o indicates negative interaction. The coefficients measure the strength of the interactions.

of the interaction (see Figure 5.1.3). In the context of biodiversity conservation, once again, what is of most interest is the persistence of a community over time, that is, the avoidance of extinction. This brings us to one of the most interesting – and one of the most controversial – questions of ecology: the relationship between diversity and stability. A deeply rooted intuition among ecologists has been that diversity begets stability. If this claim is true, it has significant consequences for conservation. For instance, it may provide a genuine rationale for the ill-fated rivet argument.[15] The loss of every species in a community must reduce its diversity to some extent (no matter how "diversity" is defined – see below). The putative diversity–stability relationship can then be used to say that it thereby reduces its stability, thus making further extinctions more likely than before. Thus, if conservation is an end, attempts should be geared toward avoiding the extinction of every single species.[16]

[15] See Chapter 1, § 1.3.

[16] As Goodman (1975), p. 261, puts it:

> The diversity-stability hypothesis has been trotted out time and time again as an argument for various preservationist and environmentalist policies. It has seemed to offer

What confuses this question, from the very beginning, is the multiplicity of possible definitions of "diversity" and "stability." For instance, a reasonable first attempt to define diversity would be to equate the diversity of a community to the number of species in it, that is, its species "richness." The trouble is that there is ample reason to doubt that richness captures all that is relevant about diversity, whether or not we are interested in only its relationship to stability. Consider two communities, the first consisting of 50 percent species A and 50 percent species B, and a second consisting of 99.9 percent species A and 0.1 percent species B. Both communities have the same richness because they both have two species; however, there is a clear sense in which the first is more diverse – or less homogeneous – than the second. Moreover, the difference is likely to be relevant. If diversity does beget stability in these communities, then that stability must be a result some interaction between the two species. If species B comprises only 0.1 percent of the community, the scope for such interaction is much less than if it comprises 50 percent. Diversity must mean more than richness. Unfortunately, though there have been several attempts to define and quantify diversity beyond richness – one of them is described in Box 5.1.2 – there has been little success in tying these concepts to theoretical rules or even to empirical generalizations.[17] (How biodiversity will be *operationally* defined in the context of its conservation will be discussed in Chapter 6, § 6.5.)

Stability turns out to be even more difficult to define.[18] At one extreme, we can try to capture stability by requiring that a community be truly at equilibrium: it does not change in composition or in the level of interaction. At the practical level, this definition faces the problem of vacuous scope: almost no natural community satisfies such a strict requirement of equilibrium. Moreover, almost every community experiences significant disturbances. With this in mind, stability has

> an easy way to refute the charge that these policies represent nothing more than the subjective preferences of some minority constituencies. . . . From a practical standpoint, the diversity-stability hypothesis is not really necessary; even if the hypothesis is completely false it remains logically possible – and, on the best available evidence, very likely – that disruption of the patterns of evolved interaction in natural communities will have untoward, and occasionally catastrophic, consequences. In other words, though the hypothesis may be false, the policies it promotes are prudent.

The rationale for biodiversity conservation developed in the last chapter does not even require that Goodman's prudential argument be sound, although it probably is.

[17] For a survey of definitions and measures of diversity, see Magurran (1988).

[18] For a very important early analysis, see Lewontin (1969).

BOX 5.1.2 MEASURES OF DIVERSITY

Ecologists often distinguish three concepts of biodiversity, usually using species as the appropriate unit to measure diversity: (i) "α-diversity," the diversity within a community/place (MacArthur 1965); (ii) "β-diversity," the diversity between communities/places (Whittaker 1972, 1975); and (iii) "γ-diversity," the diversity between regions, that is, β-diversity on a larger spatial scale (Whittaker 1972).

Though many measures of α-diversity have been proposed over the years, MacArthur's (1965) proposal to use the Shannon measure of information content in a communication process (Shannon 1948) has remained the most popular (though it is not entirely uncontroversial). According to this measure, the α-diversity of a community with n species is given by

$$\alpha = -\sum_{i=1}^{n} p_i \ln p_i = 0,$$

where p_i is the frequency of the ith species. This is a measure of the diversity of a community in the same way that the Shannon measure of information content is a measure of the variability in a signal.

Turning to the two communities discussed in the text, a simple calculation shows that the diversity of the first is given by $\alpha = 0.693$, while the diversity of the second is given by $\alpha = 0.008$, verifying our intuition that the first is more diverse than the second.

β-diversity, assuming that it can be quantified adequately, is particularly important when prioritizing places for conservation because it seeks to measure the difference between places, and the more different places are, the higher the total biodiversity content. (See Chapter 6, § 6.5, for a discussion of defining "biodiversity.")

been variously explicated using a system's response to disturbances or its tendency not to change beyond specified limits even in the absence of disturbance. Box 5.1.3 lists some of the definitions of stability that have been in vogue and how they may be measured in the field.[19]

[19] Grimm and Wissel (1997) provide a more comprehensive inventory, organized much like Table 5.1.2. This table is partly based on James Justus's dissertation, in progress ("The Stability-Diversity-Complexity Debate of Community Ecology: A Philosophical Analysis," Department of Philosophy, University of Texas at Austin).

BOX 5.1.3 DEFINITIONS OF STABILITY

	Category	Definition	Measure
	Resilience	Rate at which a system returns to a reference state or dynamic after a perturbation.[a]	The inverse of the time taken for the effects of a perturbation (e.g., of species' abundances or densities) to decay relative to the initial size.
Perturbation-based Categories	Resistance	Inverse of the magnitude of the change in a system, relative to a reference state or dynamic after a perturbation.[b]	1. Inverse of the change of species' densities or abundances relative to the original state. 2. Change of species composition relative to the original composition.
	Perturbation Tolerance/ Domain of Attraction	Size of the perturbation a system can sustain and return to a reference state or dynamic (irrespective of time taken).[c]	Perturbation size measured in natural units (perturbation may be biotic or abiotic).
	Persistence	Ability of the system to continue in a reference state or dynamic.[d]	1. The time a system sustains specified minimum population levels, e.g., nonextinction of a proportion of its species. 2. The time a system will sustain specified species compositions.
Perturbation-independent Categories	Constancy	Inverse of the variability of a system (community or population).[e]	The inverse of the size of fluctuations of some parameter of the system, such as species' richness, size, or biomass abundance.

State Change Tolerance	Probability of a return to a reference state or dynamic after a change in the value of a system's state variables.[f]	Measured in the same parameters as the state variables.
Reliability	Probability that a system (community or population) will continue "functioning."[g]	Measured in terms of how faithfully and efficiently a system processes energy and materials and engages in other biogeochemical activities.

[a] Pimm (1991).
[b] Ibid.
[c] Pianka (2000).
[d] Holling (1973); Pimm (1991).
[e] Pimm (1991).
[f] May (1973) uses a concept of stability that is very close to this: the probability of return to equilibrium.
[g] Lehman and Tilman (2000) attribute this to Naeem (1998), who, however, does not explicitly suggest the use of this criterion as a definition of stability.

How do any of these measures of stability relate to diversity? The only honest answer is that no one is sure. If diversity is interpreted as richness, it has traditionally been assumed that diversity is positively correlated with at least persistence. However, there was never much hard evidence supporting this assumption. Nevertheless, this assumption is almost certainly one of the sources of worries about the potential collapse of the biosphere due to the loss of species; thus it provides a rationale for the rivet argument, as was mentioned earlier. If stability is interpreted as a return to equilibrium, mathematical models that should answer questions about stability are easy to construct but hard to analyze unless the system is already close to equilibrium. This is called local stability analysis. The most systematic analyses performed so far give no straightforward positive correlation.[20] It was once believed that natural ecosystems are usually at equilibrium (historically

[20] See May (1973), who, however, replaces diversity with complexity in his theoretical analyses. However, his conclusions remain correct if diversity (as he measures it) is used instead of complexity.

called the "balance of nature"). But ample empirical data now suggests that his assumption is almost never correct: natural ecosystems are usually far from equilibrium.[21] Moreover, if natural selection between species occurs during the transition to equilibrium, equilibrium communities will be less rich than those that have yet to reach equilibrium. Selection between species that utilize the same resources (that is, occupy the same "niche") will lead to the exclusion of the less fit by the more fit through "competitive exclusion."[22] The eventual equilibrium community, in which selection is no longer taking place, the so-called climax community, is invariably less rich than those that temporally preceded it.

As noted earlier, the traditional assumption of a general positive correlation between diversity (as richness) and stability has been seriously challenged on both theoretical and empirical grounds since the 1970s.[23] More recently, however, Tilman and several collaborators have suggested an empirical connection between richness and stability, interpreted as constancy, in grassland habitats.[24] The scope of this generalization, even whether it can be replicated for other grasslands than those that Tilman studied, remains to be investigated. Meanwhile, Pfisterer and Schmid have produced equally compelling empirical evidence that richness is inversely correlated with stability, interpreted as resilience and resistance.[25] Much remains to be found out. Perhaps all that is certain is that McCann's confident 2000 verdict in favor of a positive diversity–stability relationship was premature.[26]

No discussion of community ecology in the context of biodiversity conservation would be complete without some mention of the theory of island biogeography and the controversies surrounding its relevance for the design of networks of biological reserves. The motivation for this theory is the species–area relation: larger areas of the same habitat type usually contain more species than smaller ones. Thus there is at least a monotonic qualitative positive relationship between species

[21] Pimm (1991) summarizes the current situation.

[22] Gause (1934) introduced this important argument; however, the range of its applicability remains controversial.

[23] See May (1973) for a theoretical survey, and Pimm (1991) for a review of the empirical work.

[24] See Lehman and Tilman (2000) for a recent summary.

[25] Pfisterer and Schmid (2002); see Naeem (2002) for a commentary.

[26] Unfortunately, McCann (2000) made this claim as if it were well established, in a special section of *Nature* devoted to biodiversity conservation. Because of *Nature*'s prominence as a journal, McCann's claim has the potential to mislead a large section of the scientific community.

richness and area. But what is the form of this relationship? Moreover, what is the mechanism responsible for it? In spite of sporadic work over almost an entire century, these remain open questions. Perhaps the most popular answer to the first question is a power law going back to Olof Arrhenius:[27] $S = cA^z$, where S is the number of species, A the area, and c and z constants. This power law represents what is often called the "species–area curve." Turning to the question of mechanisms, traditionally, the species–area relation was attributed to environmental heterogeneity. Larger areas were presumed to have greater habitat heterogeneity and could, therefore, host a larger number of species, each with its own specific needs. In recent years, the relation is more often attributed to the belief that larger areas can support larger populations of any species.[28] Thus fewer populations are likely to become extinct in a larger area than in a smaller one during any given period of time. Consequently, on the average, more species are likely to be present in larger areas than in smaller ones, even if both started with the same species richness.

Whether the species–area curve (rather than the mere qualitative relation) has much empirical or any theoretical support remains a matter of contention.[29] The self-educated mathematical ecologist F. W. Preston was a strong advocate of the power law model, which he believed to be the result of a dynamic equilibrium of the exchange of species between isolated habitat patches.[30] (The same idea had been worked out in some detail much earlier by Munroe but did not receive any attention).[31] Preston's ideas were extended by MacArthur and Wilson to construct the theory of island biogeography.[32] According to this theory, the number of species in islands with the same habitat (at the same latitude) depends only on the size of the island and its isolation. There is a dynamic equilibrium in the sense that this number does not change over time, though there is a turnover of species that changes the composition of the community. The equilibrium is supposed to be a result of a balance between immigration and extinction. The immigration rate varies inversely with the degree of isolation, while the extinction rate

[27] Arrhenius (1921); see also Gleason (1922).

[28] Margules, Higgs, and Rafe (1982) emphasize this shift in the explanation of the species-area relation.

[29] See, for example, Connor and McCoy (1979) and the reply by Sugihara (1981).

[30] Preston (1962a, b). Historical material on Preston is from Mann and Plummer (1995).

[31] Munroe (1948); Brown and Lomolino (1989) provide a discussion of Munroe's work.

[32] MacArthur and Wilson (1963, 1967).

decreases with area. Thus this theory incorporates the second mechanism for the species–area relation mentioned in the last paragraph. While some initial experimental evidence seemed to support the theory, by the mid-1970s its status had become controversial.[33]

Nevertheless, in the 1970s island biogeography began to be viewed as a model for biological reserves, which, by being surrounded by developed lands, were supposed to be similar to islands. The initially prevalent view, based on island biogeography theory, was that reserves should be as large as possible. In particular, one conclusion drawn from island biogeography theory was that "[i]n cases where one large area is infeasible, it must be realized that several smaller ones, adding up to the same total area as the single large one, are not biogeographically equal to it: they will tend to support a smaller species total."[34] Though this theory-based conclusion was even incorporated into the design of the World Conservation Strategy of the International Union for the Conservation of Nature, there was little data that supported it.[35] It was also challenged by Simberloff and Abele, on both theoretical and empirical grounds.[36] Among other things, they pointed out that several small reserves may increase the probability of the survival of species in the face of some stochastic factor such as infectious disease. This objection sparked the SLOSS (Single Large or Several Small) debate about the design of biological reserve networks. The SLOSS debate dominated discussions of reserve network design for about a decade. Meanwhile, the species–area curve itself received serious criticism. Soulé, Wilcox, and Holby predicted in 1979 from a model based on the species–area curve that the famous Serengeti National Park of Tanzania would lose 50 percent of its large mammals (15 ungulate species) in 250 years.[37] However, once Western and Ssemakula incorporated habitat diversity data in 1981, it appeared that only one such species was doomed for extinction.[38] There are many other such examples, and it is hard not to sympathize with Zimmerman and Bierregard,

[33] See, for instance, Gilbert (1980); problems with the assumptions of the theory and the difficulty of calibrating its parameters were pointed out as early as 1969 in Sauer's (1969) review of MacArthur and Wilson's book.

[34] May (1975), p. 177. See also Wilson and Willis (1975), Diamond (1975b, 1976), Diamond and May (1976), and Terborgh (1975, 1976) for a sample of the consensus that was achieved.

[35] International Union for the Conservation of Nature (1980).

[36] Simberloff and Abele (1976a); for responses, see Diamond (1976), Terborgh (1976), and Whitcomb, Lynch, and Opler (1976). Simberloff and Abele (1976b) contains their reply.

[37] Soulé, Wilcox, and Holby (1979).

[38] Western and Ssemakula (1981).

who observe that, except for the ecological truism that species richness increases with area, there is little of value that the species–area curve and the theory of island biogeography contribute to reserve design.[39] Luckily, the putative relevance of the species–area curve and island biogeography for biodiversity conservation was never fully accepted. Important early criticism of the use of island biogeography theory for reserve network design came from Margules, Higgs, and Rafe in 1982.[40] They pointed out that the theory was yet to be empirically established in the field; that biological reserves were not very similar to islands, because spaces between reserves were not completely uninhabitable by the species in the reserves (unlike the case of oceans separating islands); that habitats are largely heterogeneous rather than homogeneous (as assumed in the theory); and that species richness should not be the only criterion used to select reserves. By 1986, it had become clear that there would be no winner in the SLOSS debate; since then, there has been no unequivocal role for island biogeography theory to play in the design of biodiversity reserve networks.[41]

However, the SLOSS debate has left behind one important though equally contentious legacy that remains relevant to biodiversity conservation today, namely, the contemporary debate over the effects of habitat fragmentation. In the fragmentation debate, unlike the SLOSS debate, it is not assumed that the landscape "matrix" between habitat fragments is entirely inhospitable to taxa present in the habitats – thus the fragmentation question cannot be dismissed as easily as the SLOSS question eventually was. The unanswered question is whether fragmentation poses a serious concern for the persistence of biodiversity. Fragmentation is almost always accompanied by (and usually caused by) habitat loss.[42] What is at stake is whether fragmentation per se, rather than habitat loss, is responsible for the well-documented loss of biodiversity (usually interpreted as species richness) when there

[39] Zimmerman and Bierregard (1986).

[40] Margules, Higgs, and Rafe (1982). Earlier, Higgs (1981) provided a theoretical analysis comparing a single large reserve to two smaller ones, each of which had half the area of the first. Whether the single reserve or the pair permitted the coexistence of more species depended on the context; the island biogeography theory did not provide unequivocal results. Rosenzweig (2005) reaches the same conclusion after a critical review of more recent work.

[41] Soulé and Simberloff (1986). The history of the SLOSS debate has yet to receive the philosophical attention it deserves (see, however, Shrader-Frechette 1990 and, especially, Kingsland 2002). Mann and Plummer (1995) include a useful popular history.

[42] Fragmentation typically occurs because of anthropogenic conversion of the habitat between the fragments into "nonhabitats" for the taxa in question. See Fahrig (2003) for a recent review.

is fragmentation accompanied by habitat loss. The answer is far from clear; it has even been suggested – on the basis of some empirical studies – that fragmentation may increase biodiversity.[43]

(iii) *Ecosystem models*: the term "ecosystem" was coined in 1935 by the plant ecologist A. G. Tansley, who defined it as "the whole *system* (in the sense of physics) including not only the organism-complex [that is, the community], but also the whole complex of physical factors forming what we call the environment of the biome – the habitat factors in the widest sense."[44] Tansley went on to argue that ecosystems "are the basic units of nature on the face of the earth." For Tansley, using the term "ecosystem" implied a physical description of a community in its habitat; though that perspective still illuminates ecosystem studies to some extent (see below), it is no longer a necessary or even common connotation of the term "ecosystem" today.

The introduction and rapidly growing popularity of the term "ecosystem," especially during the late 1950s and 1960s, was marked by two major cognitive and one sociological shift in the practice of ecology. (a) Because it came at the end of the so-called golden age of theoretical population ecology of the late 1920s and 1930s,[45] the turn to ecosystems helped to shift the emphasis from populations with interacting individuals to much larger and more inclusive systems. In this sense, it was a deliberate anti-"reductionist" move.[46] Ecosystem enthusiasts, by and large, follow a long holistic tradition in natural history that tends to deify complexity and to deny the possibility of explaining wholes in terms of their parts.[47] (b) A second cognitive shift is that ecosystem studies involve models based, at least in part, on non-biological variables.[48] For instance, instead of tracking individuals or even species in communities, models may track energy or matter flow in food webs as a whole. (c) At the sociological level, the expansion of ecosystem studies led to what one historian has called the invention of "big biology" in the 1960s, chiefly in the United States.[49]

[43] Of seventeen studies summarized by Fahrig (2003, Table 2), four showed no effect, one showed a negative effect, seven showed a positive effect, and the rest gave mixed results.

[44] See Tansley (1935); emphasis in the original.

[45] Scudo and Ziegler (1978) categorize this period as the "golden age."

[46] Reductionism is being used here in the sense of Sarkar (1998a); it will be further discussed in the next section.

[47] See McIntosh (1985), Chapter 6, for more details on this point.

[48] Golley (1993) has traced the history of the "ecosystem concept," including both the philosophical and the scientific developments discussed in this section, in great detail.

[49] See McIntosh (1985), p. 205.

These studies – for instance, the massive Hubbard Brook Ecosystem Study[50] – required more than just many biologists working together. They also demanded that other specialists, including geochemists and soil scientists, be brought in so that all of the relevant physical parameters of ecosystems, not just the biological ones, could be tracked simultaneously. This was biologists' attempt to follow the practice of publicly funded Big Science, initiated by physicists during the Manhattan Project and subsequently profitably exploited by social scientists, beginning in the 1950s.

Until just a decade ago – some sixty years after the introduction of the term "ecosystem" and thirty years after a veritable explosion of ecosystem studies – it was less than clear what important new insights this disciplinary move had produced. The trouble was that, at this level of analysis, very few general claims could be sustained. Those that could – for instance, the claim that the sun is ultimately the source of all energy in biological systems or that primary producers have to contain chlorophyll or some other such molecule – were usually trivial and well known long before the initiation of systematic large-scale ecosystem studies in the 1960s. Usually ecosystem studies produced detailed analyses of the nutritional or climatic requirements of particular communities. But the details of nutritional requirements were either so general as to be of no great interest, or so specific that they were rarely transportable from one ecosystem to another. Almost all of what is known about the climatic requirements of vegetation types (and of other communities) was known to biogeographers long before the invention of ecosystem studies. Even the carbon and nitrogen cycles had been worked out long before the advent of ecosystem studies as an organized discipline.

Nevertheless, the physical characteristics of habitats do matter to the organisms living in them. Moreover, physical changes on a global scale – for instance, climate change through global warming – have serious long-term implications for biodiversity conservation and cannot be ignored. These other changes include the increasing concentration of carbon dioxide in the atmosphere and changes in the biogeochemistry of the global nitrogen cycle, as well as changes in land cover and land use.[51] During the last decade, ecosystem studies and models have

[50] See, for example, Bormann and Likens (1979a, b). Both McIntosh (1985) and Golley (1993) identify this study as critical to the attempt to establish ecosystem studies institutionally.

[51] See Vitousek (1994) for a survey.

finally matured to produce not only novel intellectual insights about ecosystem processes, such as the carbon and nitrogen cycles, but results that are increasingly relevant to biodiversity conservation.[52] For instance, the effects of disturbance and fire on ecosystem processes are now being seriously assessed.[53] Consider, as an example, one of the more interesting analyses that shows how ecosystem modeling may become highly relevant to biodiversity conservation. Ryan has used a complex model tying physiological processes to the physical environment to suggest that increased temperature will make maintenance respiration (which represents the physiological costs of protein synthesis and replacement, membrane repair, and the maintenance of ion gradients in cells) for plants more difficult.[54] This is important because the total plant respiration, including maintenance respiration, is an important component of the carbon balance in any ecosystem. Ryan's model is based on observed statistical associations of the different parameters; the underlying mechanisms resulting in the high sensitivity of maintenance respiration to temperature change (as well as changes in many other physical parameters, such as carbon dioxide and protein concentrations) remain unknown. Ryan's result is troubling because of the ongoing climate change through global warming. In another example, Aerts and Chapin have provided a systematic review of the nutritional ecology of wild plants, including nutrient-limited growth, nutrient acquisition, use efficiency, and recycling through decomposition.[55] This review underscores the conclusion not only that plant growth in terrestrial ecosystems is very often controlled by nitrogen availability in the nutrients, but also that it is often similarly dependent on phosphorus availability. There are many other similarly important recent examples.

What has made much of the new work possible is not only increased experience with ecosystems but also significant technical innovation, including the advent of high-speed microcomputers, satellite imagery, and Geographic Information Systems (GIS), which will be discussed next (in § 5.2). The future of ecosystem ecology appears much more secure today than it did a decade ago. Moreover, recent developments have underscored its importance for biodiversity conservation.

[52] On such models, see, especially, Sellers et al. (1997). Prentice et al. (2001) summarize the current situation.

[53] See, for instance, Hobbs and Huenneke (1992) and Kasischke, Christensen, and Stocks (1995).

[54] Ryan (1991).

[55] Aerts and Chapin III (2000).

5.2. NEW DIRECTIONS

The use of scientific ecology in devising strategies for conserving biodiversity in the field is made particularly difficult by four (partially related) epistemological problems of which ecologists have long been aware.[56] (i) The *complexity* problem: systems in the field are far more complex than those that can be modeled or even experimentally studied in the laboratory. Sometimes this complexity precludes the design of field experiments with adequate controls. (ii) The *uniqueness* problem: some systems are so unique in their structure or composition that laboratory experiments cannot be used to derive insights about them. Once again, uniqueness may also make the design of controlled experiments difficult, if not impossible. (iii) The *estimation* problem: even the simplest of parameters used in ecological models – for instance, the intrinsic growth rate and carrying capacity in models in population ecology – are often impossible to estimate accurately in the field.[57] Estimation of the strengths of interactions between species in models in community ecology is often even more intractable. (iv) The *structural uncertainty* problem: slight changes in the assumptions of the models may lead to radical changes in the predictions made using them.[58] This results in structural uncertainty in our knowledge of the models (which will be further discussed in Chapter 7, § 7.4). Complexity aggravates the problem of structural uncertainty. What makes this situation particularly troubling in the context of biodiversity conservation is that decisions with potentially irreversible effects must be made in spite of these difficulties.

Nevertheless, there have been two recent developments that augur well for the future use of scientific ecology in biodiversity conservation, insofar as they often help to mitigate the complexity and uniqueness problems and have the potential to mitigate the structural uncertainty problem as well. Both developments have been made possible only by the astronomical increase in the speed and ease of computation since the early 1980s:

(i) In the models of population ecology discussed earlier (in § 5.1), the first option mentioned was that populations are characterized by their *state* variables, parameters such as size and density describing the

[56] See, for example, Peters (1991) and Shrader-Frechette and McCoy (1993). Note that these problems are not logically independent of each other.

[57] This is part of the more general partial observability problem of ecological systems – see Chapter 7, § 7.4.

[58] Levins (1966) discusses ways to make reliable inferences from models in such a situation; for further discussion, see Wimsatt (1987). Orzack and Sober (1993) criticize Levins's arguments; for a response, see Levins (1993). For more on structural uncertainty, see Chapter 7, § 7.4.

population as a whole and – with two exceptions – ignoring individual differences. (The exceptions are age and stage; the age or stage structure of populations [the fraction of individuals in each age or developmental stage group] is sometimes incorporated into the traditional models of population ecology.) Around 1980, that situation began to change, and the second option of characterizing populations by individual variables began to be systematically pursued. This shift has led to the emergence of "individual-based models" (IBMs) that incorporate individual differences explicitly.[59] These models have begun to provide some insight into the questions that require answers in the context of biodiversity conservation. IBMs represent a population as a collection of individuals with variable properties – size, growth rate, biomass, and so on. The interactions between individuals are incorporated into the model. Since, because of their sheer complexity, such models are usually impossible to study analytically, they are studied by simulation on a computer. The wealth of detail that can be incorporated into IBMs allows specific predictions to be made both in general, mitigating the complexity problem, and also for very specific values of the parameters, mitigating the uniqueness problem. Moreover, IBMs can be run on varying sets of parameters, thus potentially estimating the extent of structural uncertainty of a model. Part of the attraction of IBMs has been their relative predictive success compared to other types of ecological models.[60]

In the context of environmental studies (including that of biodiversity), IBMs are particularly useful because they can also be spatially explicit, that is, they can incorporate locational relationships between the individuals being modeled. However, because of computational complexity, the size of the individual areas for which these models can be effectively simulated is highly restricted, usually to about 10^6 sq. m. Nevertheless, these models have even been used to assess change on a global scale. For instance, forest models (which are among the most successful IBMs) have been used to assess the effect of climate change on the atmosphere because of a potential for the breakdown of the presumed balance between production and decomposition of carbon-containing compounds. Such an extrapolation of scale relies

[59] See the important review by Huston, DeAngelis, and Post (1988), which correctly predicted that IBMs would change the nature of ecological modeling.

[60] Shugart (1984); Shugart, Smith, and Post (1992). While these may be uncharacteristic in the sense that they are unusually predictively successful, they provide a standard to which all IBMs may aspire.

on sampling each of the terrestrial life-zones and constructing some IBMs for all of them, and subsequently integrating the results.[61] The future will show how reliable this strategy is. Recently, IBMs have also begun to be used for population viability analysis, tracking the trajectory of each individual during its lifetime.[62] This use is likely to grow. In both the situations discussed here, the main problem with the use of IBMs is the immense quantity of reliable data that they require.

Within the context of population ecology, because the behavior of the entire population is putatively being explained on the basis of the properties of the individuals within them, using IBMs is, philosophically, a reductionist strategy called "methodological individualism."[63] Here, "reductionism" means that models of large systems should try to explain their behavior *entirely* in terms of the properties of their parts (nowhere referring to intrinsically "systemic" properties). More specifically, such a reductionism amounts to the assumption that properties and interactions of individuals alone suffice to explain all behavior at the level of populations (and higher units): there is no need for explanations to refer to higher-level or systemic properties that cannot be defined in terms of individual properties (for instance, the density of a population). Moreover, because interactions between individuals of different species can also be incorporated into these models, community-level properties can potentially be explained by IBMs. For instance, the structure of food webs can potentially be explained by IBMs that take into account habitat size and resources.[64] Thus, even community structure is potentially reducible to individual interactions. In this sense, community ecology, like population ecology, is also being reduced to IBMs.[65] In this odd way, IBMs are unifying at least these two subfields within ecology. It remains surprising how little philosophical attention IBMs have so far received. If these succeed, they will help to end the long, and arguably unfruitful, tradition of holism in ecology.

[61] See Shugart, Smith, and Post (1992) for details of the complex integration procedure that must be followed.

[62] See Beissinger (2002) and Beissinger and Westphal (1998). Both are optimistic about the future of IBMs in PVA.

[63] For more on reductionism, see Sarkar (1998a).

[64] For instance, Spencer (1997) constructs such a model for small freshwater benthic habitats with three trophic levels (algae, herbivorous invertebrates, and predatory invertebrates).

[65] Fryxell and Lundberg (1998) take this unificatory strategy even further to include some evolutionary considerations.

(ii) What has most irrevocably altered the practice of biodiversity conservation and, at least arguably, is now beginning to alter the practice of ecology is the advent of Geographic Information Systems (GIS), which allow the detailed spatial representation and rapid manipulation of georeferenced data on computers. (All of the examples presented in Chapter 6 are based on GIS models.) GIS originated in sparsely populated Canada, which, until the 1950s, viewed land and other resources as unlimited. The late but inevitable realization that this was not the case led the Canadian federal government to initiate a national inventory of land and other natural resources. The purpose of what was christened in 1963 the Canadian Geographical Information System (CGIS) was to analyze data collected by the Canada Land Inventory (CLI) to produce statistics that could be used to develop land management plans for effective resource utilization in large areas of rural Canada.[66] The CLI produced seven maps classifying land according to (a) soil capability for agriculture, (b) recreational potential, (c) habitat potential for ungulates, (d) habitat potential for waterfowl, (e) forestry potential, (f) present land use, and (g) shoreline. Constructing the CGIS meant developing techniques for the rapid handling and analysis of these maps and the data on which they were based. Today's commercial packages parasitize key conceptual and technical innovations of the CGIS. At the technical level, when the CGIS project was initiated there was no prior experience of how to structure georeferenced data internally (within the computer); there was no technique for the overlay of maps or for calculating area. An experimental scanner to scan map data had yet to be built.

Among the conceptual innovations, the most important was the distinction between (a) the data used to draw the polygons forming the boundary of a place (locational information) and (b) the set of features the place has, that is, its attributes. Polygons need not have the same size or geometry. When ecological populations and communities are modeled on a GIS, explicit asymmetric irregular spatial information can be incorporated without unrealistic simplifying assumptions, such as that of representing the entire spatial structure as a square or some other such regular geometric grid. The exploitation of this possibility takes spatially explicit ecological modeling beyond its traditional confines

[66] Roger Tomlinson, who directed the project, and Duane Marble are the two individuals usually credited with inventing the acronym "GIS." Much of the history recounted in the text is from Tomlinson (1988).

in which the only spatial structures that could be represented were those with symmetric geometric structure. Though GIS-based ecological modeling is still in its infancy (an early example will be discussed in the next paragraph), it is clear that these techniques will allow the construction of spatially explicit ecological models at a level of detail that was impossible before. Moreover, IBMs can now be constructed with such detailed spatial representation. The confluence of IBMs and GIS is arguably the most promising area of ecological modeling today.

Just as important as the distinction between polygon and attribute was the decision for the "vectorization" of scanned images. Scanned images gave "raster" data, that is, data in the form of regular grid points that either do or do not possess a specific property, for instance, the presence of a given vegetation type. Vectorization is the replacement of these point-based structures with lines that are naturally interpreted, such as boundaries of habitat types. What is critical is that these lines can then be joined to form polygons. Raster data can be obtained from a variety of sources, including maps and photographs; in the present context what is critical is that raster data can be obtained by remote sensing through satellite imagery, from which the distribution of many vegetation and soil types can be inferred. As early as 1989, Running and several collaborators estimated the annual evapotranspiration and net photosynthesis for a 28 × 55 sq. km region of Montana using a GIS software package.[67] The study region was divided into 1.1 × 1.1 sq. km cells defined by satellite sensor pixel size. A GIS package was used to integrate topographic, soil, vegetation, and climatic data. Ecological assumptions were then used to construct models that were used to predict evapotranspiration and net photosynthesis. The results obtained were in fairly good agreement with experimental data. In Chapter 6 (§ 6.1) it will be pointed out how such data can be critical to systematic conservation planning, even though this process is subject to many pitfalls.

Within ecology (and conservation biology), the use of GIS-based models is the analog of visual modeling in other sciences.[68] It is no longer controversial that visual representation, at least as a heuristic,

[67] Running et al. (1989) provide what is probably the first complex example of GIS-based spatial ecology.

[68] Spatial modeling on regular geometric grids is not analogous to visual modeling in the same way. Proper visual representation is supposed to be as veridical as possible. This means that as much spatial detail as possible should be incorporated into a model. Representing spatial structure as having a global regular geometry, as traditional spatially explicit ecological models have been forced to do (because of the complexity otherwise encountered), goes against the spirit of this veridicality requirement.

offers resources for scientific innovation not offered by purely linear representations (such as linguistic or mathematical representations). GIS-based models constitute two-dimensional visual representations of ecological systems. It is likely that these representations incorporate spatial insights that will result in new and fecund directions for ecological modeling.

Nevertheless, what is somewhat philosophically troubling about the use of GIS in ecology is the conceptualization and representation of geographical information as both (a) a linked set of places, linked in the sense that the places must maintain fully precise adjacency relations, and (b) an unlinked set of attributes (for instance, the presence or absence of species or other biological features). There is something disarmingly natural about this: it does seem to capture the geographical rootedness in place that lies at the basis of planning for biodiversity conservation. But this choice of representation has its costs: the mode of representation that is at the core of GIS makes it "natural" to represent systems in such a way that certain types of relationships tend to be lost, or at least to be relegated to the background, while others receive emphasis. Consider the following example. Carnivore species cannot be present at any place unless prey species also exist. This trivial and obvious ecological fact cannot be explicitly represented using the standard resources of any GIS package (that is, it cannot be represented without writing special programs). Attributes are represented without relations between them. This encourages, though it does not require, analyses that do not use relations between attributes. (Obviously, one can start with a GIS-based representation and add other relations as part of the superstructure of the model.)[69] Philosophers of science have long known that modes of representation influence the introduction and development of conceptual systems based on them. GIS may have such an influence through representational choices that guide ecology down a path where relations between attributes receive less emphasis than they would in traditional ecological models.

5.3. EXTINCTION

To conserve biodiversity is to ward off extinction. If there is any aspect of biodiversity conservation on which scientific ecology should shed

[69] Some GIS software packages, such as ARC/INFO (see, e.g., Zeiler [1997]), explicitly facilitate the addition of such superstructures.

significant insight, it is extinction. Scientific ecology should be able to provide

(i) a relatively accurate and complete list of potential factors responsible for extinction;
(ii) reliable estimates of the current and background extinction rates (that is, the average extinction rate through evolutionary history, except during the mass extinctions);
(iii) an assessment of whether the current extinction rate is sufficiently higher (if indeed it is higher) than the background rate to warrant talk of a "crisis"; and
(iv) an indication of whether the increase in the extinction rate (if any) is anthropogenic.

The way in which these questions have been phrased recognizes that extinction is a "normal" part of evolution, "normal" in the sense that extinctions have always occurred during the course of biological evolution on Earth.

It is uncontroversial that ecology does show that many recent extinctions are anthropogenic, though it is perhaps a conceptually nontrivial extrapolation from this fact to conclude that they are "unnatural." After all, *Homo sapiens* is also a biological species, and many "natural" extinctions can be attributed to the action of one species on another. This is a philosophically interesting question that has no simple answer; luckily, the normative rationale for conserving biodiversity developed in this book (Chapter 4) does not require an answer to it. However, even if most extinctions that are currently taking place are anthropogenic, a comparison of present and background extinction rates alone does not provide any reliable evidence for that fact. The reason for this is that extinctions have not been occurring at a constant rate through all of evolutionary history. Extinction has always been an episodic process. Most paleontologists recognize about fifteen major episodes during which extinction rates were much higher than the background; five of these are usually distinguished as being much more significant than the others and are called "mass extinctions." But even at other times, the extinction rate has varied. Consequently, even if the present extinction rate is significantly higher than the background, this is *by itself* no indication that, from an evolutionary point of view, something unusual is going on. We may still wish to interfere with extinctions. But we would then have no special reason to worry about the potential fate of biodiversity on Earth because of some impending collapse.

The question of the relation of current extinction rates to the background is related to that of whether there is a biodiversity "crisis." It is impossible to say anything sensible about the latter question unless the term "crisis" is precisely defined. To attempt such a definition seems to be an utterly thankless task given the vagueness of the term in the present context. Instead, the best solution may be simply to adopt the convention of saying that there is such a crisis only if extinction rates begin to match those of any of the mass extinctions.[70] This is a restrictive choice; it is at least arguable that it would be equally reasonable to use all major extinctions as the standard of reference. However, as will be argued later, prudence dictates a restrictive choice. This way of talking has the virtue that reference to "crisis" is restricted to a situation where it is uncontroversial that there is a very significant loss of diversity (these are "mass" extinctions). There are many biodiversity conservationists who will probably not endorse such a restrictive proposal – because, as will be argued, it is far from obvious that a mass extinction is going on. Recall, from the discussion at the end of Chapter 1 (§ 1.3), that "crisis" has a rhetorical power that some biodiversity conservationists may well wish to deploy. Nevertheless, as noted there, such a use of rhetoric may bring with it the undesirable prospect of a loss of public credibility, which may have a long-term deleterious effect on the adoption of public policies designed to ensure biodiversity conservation. A restrictive choice should help to disarm critics of biodiversity conservation in the public arena. Prudence suggests the adoption of the restrictive definition of "crisis" offered here. There are at least some biodiversity conservationists who now accept the force of this prudential argument. A recent editorial in *Conservation Biology* argued that "the oft-repeated definition of conservation biology as a crisis discipline has outlived its usefulness."[71]

What makes discussions of the extinction rate most difficult is the familiar problem of uncertainty. The fossil record is the source of all that is known about extinctions over geological time. This already introduces a significant bias. Not all species fossilize equally well; those with soft bodies do not fossilize at all. The species that have had the highest probability of preservation are marine species with highly mineralized exoskeletons. Around 95 percent of fossil species are marine animals.[72] Being a member

[70] As will be shown later in this section, this amounts to saying that the current extinction rate is at least twice the background rate.

[71] Redford and Sanjayan (2003), p. 1473; they go on to argue that "the founding vision, tone, and language of conservation biology, so crucial in the infancy of the field, is now casting long shadows.... Our focus on crisis has hampered conservation biology in achieving a scale of action required to match the world's environmental problems. Despite our best efforts to launch our cause into mainstream culture, the world is suffering from crisis fatigue" (p. 1473).

[72] Raup (1976).

of a marine species increases the probability of fossilization because, after death, such an individual will have been deposited on the ocean floor, where sedimentary rocks are gradually formed. These rocks are the main source of fossils. A durable mineralized skeleton helps to prevent degradation by physical factors. Most such species were invertebrates. Given the nature of the data that are available, the discussion of this section must necessarily be based primarily on the marine invertebrate fossil record. Extrapolation from this record to other groups, particularly to plants, is obviously problematic; what is said here about patterns of extinction may have little relevance to historical patterns of extinction in plants.

The most important generalization that can be drawn from the fossil record is that extinction has been episodic and that major extinction events *may* have been periodic, with a periodicity of 26 million years.[73] As noted before, there have been five major mass extinctions, starting with the end-Ordovician, some 439 million years ago (Mya). Before this, there is some evidence that there was an episode of significant diversity loss late in the Precambrian period (about 700 Mya), but the Precambrian fossil record is far too sparse to permit any detailed analysis of the type of diversity that was lost. From the entire fossil record, it is estimated that 94–98 percent of all species that have existed are now extinct. However, 90–96 percent of all extinctions took place outside these major extinction episodes.[74]

The five mass extinctions will be considered in some detail because the question of whether we are in a biodiversity "crisis" will be interpreted as the question of whether the present extinction rate is comparable to that during at least one of these episodes.[75] There is no universally accepted single definition of "mass extinction," though Jablonski's suggestion that it is "one that doubles the extinction background level among many different kinds of plant and animal groups" seems reasonable.[76] It picks out exactly those five episodes that have conventionally been regarded as mass extinctions. It is important to reemphasize that these episodes have been selected on the basis of marine invertebrate data. These data are probably not representative of other taxa, let alone all species.

[73] See Raup and Sepkoski (1984). The periodicity hypothesis remains controversial (see Raup 1991).

[74] See May, Lawton, and Stork (1995) and Raup (1986, 1992).

[75] The numbers given here are calculated primarily from Jablonski (1995); Raup (1991) and Groombridge (1992) have also been used. Virtually every one of them, whether it be the length of an extinction episode or the percentage of taxa that were lost, is subject to ongoing controversy.

[76] See Jablonski (1993), p. 48.

Table 5.3.1. *Extinction rates during the five mass extinctions*

	Families		Genera	
	Observed Extinction (%)	Calculated Species Loss (%)	Observed Extinction (%)	Calculated Species Loss (%)
End-Ordovician 439 mya	26 ± 1.9	84 ± 7	60 ± 4.4	85 ± 3
Late Devonian 367 Mya	22 ± 1.7	79 ± 9	57 ± 3.3	83 ± 4
End-Permian 245 mya	51 ± 2.3	95 ± 2	82 ± 3.8	95 ± 2
End-Triassic 208 mya	22 ± 2.2	79 ± 9	53 ± 4.4	80 ± 4
End-Cretaceous 65 mya	16 ± 1.5	70 ± 13	47 ± 4.1	76 ± 5

Note: All data – and extrapolations – are for marine invertebrates. The first column lists the five mass extinctions; the next pair of columns give percentage-loss numbers for families and the species in them; the final pair does the same for genera. For both families and genera, the observed extinction percentages are directly estimated from the fossil record. This permits extrapolation to calculate species-loss percentages. The agreement of species-loss percentages calculated from observed family-extinction percentages to that obtained from genera-extinction percentages provides an internal consistency check on the extrapolation. This agreement is very close.
Source: Jablonski (1995), p. 26.

Details of the five mass extinctions follow in chronological order; the numerical data are summarized in Table 5.3.1.

(i) *end-Ordovician*: this mass extinction occurred about 439 Mya and seems to have been correlated with global glaciation (the Hirnantian glaciation). It was of relatively short extent, with three separate episodes spread out over 500,000 years. It is believed to be the second most severe of the five mass extinctions; 26 percent of families of marine invertebrates are believed to have become extinct. The marine data are probably more indicative of the loss of diversity in all taxa in this case than in the others, because plants and animals began to colonize land systematically only about 450 Mya.

(ii) *late Devonian*: this episode occurred about 367 Mya and was probably an extended event, though the evidence for this is uncertain. It probably consisted of several relatively short events rather than one protracted one. There is some controversial evidence that extraterrestrial impacts may have played a role in causing this event. Twenty-two percent of families of marine invertebrates are observed to have become extinct.

(iii) *end-Permian*: this occurred about 245 Mya and was the most severe of the mass extinctions. It is generally accepted that this was an extended event, taking 5–8 million years. It seems to have been accompanied by several geologically rapid physical changes, including the formation of the supercontinent Pangea, climatic changes, and volcanic activity. Extraterrestrial impacts have so far not been implicated. There was about a 51 percent decline in the number of families of marine invertebrates and a 82 percent decline in the number of such genera. Extrapolation shows that the number of species may have declined by as much as 95 percent.

(iv) *end-Triassic*: this occurred about 208 Mya. It was probably also an extended event, though the evidence for this is uncertain. There is again some controversial evidence that extraterrestrial impacts may have played a role in causing this event. Twenty-two percent of marine invertebrate families are observed to have become extinct.

(v) *end-Cretaceous*: this is the most recent mass extinction, well known to the general public because it included the extinction of the dinosaurs. It occurred about 65 Mya, and because it is so recent, it is the best documented in the fossil record. Land animals seem to have suffered a little more than marine animals.[77] Land plants seem to have done a little better than both. However, angiosperms were hard hit, and the fossil record shows a sudden temporary shift from angiosperm (flowering plant) pollen to fern spores. This extinction is believed to have been a relatively short event, and it now appears fairly certain that the cause was an asteroid impact.[78] Only 16 percent of marine invertebrate families are observed to have become extinct. As mass extinctions go, this one was minor.

Potentially, a sixth mass extinction began with the advent of, or at least with the expansion of the range of, *Homo sapiens*. It may have begun to have had serious impacts on other species as early as the late Pleistocene era (which lasted until about 10,000 years ago). The arrival of this species on new continents – Australia around 50,000 years ago and America around 11,000 years ago – seems to have coincided with large-scale extinctions in several taxa (though these data remain controversial). Australia lost almost all of its genera of very large mammals, snakes, and reptiles and about half

[77] Raup (1991).

[78] The asteroid impact model was first – and controversially – suggested by Alvarez et al. (1980). Evidence for it is steadily growing, though there are still some skeptics – see Jablonski (1993, 1995).

of its flightless bird species. North America lost about 73 percent and South America about 84 percent of its genera of large mammals.[79] If these are the taxa used, and these losses were on a global scale, a mass extinction would have begun. But the losses were not global, and other taxa almost certainly suffered much less. Nevertheless, provided that these extinctions were anthropogenic – and it may be the case that they are not; a climatic etiology remains a strong possibility – what becomes troubling is the scale of the human impact on biodiversity even before the invention of technologies that permit the almost instantaneous destruction of habitats. On the other hand, 40 percent of the genera of large mammals have become extinct in Africa during the last 100,000 years,[80] and this certainly cannot be attributed to the *recent* arrival of *Homo sapiens*. Early humans almost certainly evolved in Africa, and the rate of extinction has been relatively uniform in that continent. Moreover, human activity probably cannot be implicated in this case, because human technology remained primitive until very recently.

Turning to recorded history, the pattern of extinction since 1600 is revealing. By and large, whether mollusks, birds, or mammals are studied, both on islands and continents the extinction rates consistently increase. This is easy to document for data up to 1960. For instance, from 1600 to 1719, the recorded numbers of extinctions for mollusks, birds, and mammals are, respectively, 0, 20, and 0 for islands and 0, 0, and 0 for continents; from 1720 to 1839, the corresponding pair of triplets is (0, 24, 1) and (0, 1, 3); from 1840 to 1959, it is (115, 45, 10) and (31, 8, 17).[81] Even allowing for all the different biases in this data (uneven records at different places and different times, more attention being paid to some taxa than others, and so on), it is difficult to disagree with the claim that the extinction rate, has been increasing. After 1960, there is an apparent decline in extinction rate, but that is purely an artifact. Under the Convention on Trade in Endangered Species of Wild Fauna and Flora (CITES), it takes fifty years for a species to be listed as extinct after its last sighting. Criteria used by the International Union for the Conservation of Nature (IUCN) and the World Conservation Monitoring Centre (WCMC) to compile their well-known Red Books of extinct species and those at risk are equally stringent. Many species accepted by biologists as extinct – probably hundreds – are yet to be listed. In

[79] See Groombridge (1992), p. 198.
[80] See May, Lawton, and Stork (1995).
[81] See Groombridge (1992), p. 200.

general, extinction rates have been higher on islands than on continents.[82] In some cases, it is clear that the extinctions are anthropogenic. Polynesians first settled in the Hawaiian islands during the fourth and fifth centuries CE. They are implicated in the extinction of about 50 of the 100 endemic species of birds on those islands. Similar stories can be told about many other islands, including those of New Zealand and Madagascar. That more recent extinction rates are higher than earlier ones is at least partially explained by the increased rate of habitat destruction.

In the context of biodiversity conservation, the estimation of three types of extinction rate is critical:

(i) The *background extinction rate* (for all species or for some taxon) serves as a benchmark for assessing the severity of the current extinction rate. This rate must be estimated from the fossil record; clearly, no other alternative is available. Moreover, because of the incompleteness of the fossil record, almost all of the data comes from plant and animal records since the Cambrian (570 Mya). Using some 8,500 cohorts of fossil genera, Raup estimates 11 million years as the average life span of an invertebrate species.[83] For all species, this is one of the highest estimates, though insect species in general probably have even higher life spans. Martin puts the average life span of mammal species at 1 million years.[84] It is likely that 5–10 million years is a reasonable estimate of the average life span of a species (in general). Now suppose that 5–10 million species are currently alive. Also suppose that this number remains reasonably constant over evolutionary time. This is clearly a false assumption: the current era is at a peak of biodiversity. However, there may well be as many as 30 million species currently alive, and 5–10 million may be a good estimate of the average number of species on Earth for the last 500–600 million years. Then the background extinction rate, as estimated from the fossil record, is about one species per year. No serious meta-analysis has been attempted to judge the reliability of this final estimate (which is generally accepted). Given the uncertainties at each stage of the calculation, it is hard to suggest that the effort that would be required to perform such a meta-analysis would be worthwhile. Probably no one will be surprised if this estimate is off by a factor of 10 or even 100.

[82] This is at least partly explainable on the basis of the theory of island biogeography.
[83] Raup (1978).
[84] See Martin (1993).

(ii) The *current extinction rate* (again, for all species or for some taxon) will provide the basis for assessing the future of biodiversity so long as present trends continue. The type of data noted earlier, the recorded instances of extinction since 1600, was compiled by the IUCN and the WCMC.[85] These compilations not only list extinct species but also introduce categories referring to the severity of threat to the survival of others. These include not only "probably extinct," "endangered," and "vulnerable," in decreasing order of severity, but also categories such as "rare," which are also of potential concern. Some of the data are summarized in Table 5.3.2. This table pools together as "threatened" the IUCN categories of "probably extinct," "endangered," and "vulnerable." If these data are taken to be definitive, then two conclusions emerge. The extinction rate estimated from the historical record is significantly higher than the background rate – it is about three species per year. Nevertheless, given the uncertainties, it is not high enough to suggest that a mass extinction is going on, because for most groups the extinction rate is still not more than twice the background rate.[86] Therefore, there is no reliable basis for speaking of a biodiversity "crisis."

The trouble is that these data are far from definitive. There are at least three sets of problems:[87] (a) The data reflect the fact that some taxa have received far more attention than others. Those that have received more attention are exactly the ones for which extinction and endangerment are highest in the data. The reason why the extinction and threatened percentages are so much higher for birds and mammals than for other groups is because of such attention. As Gaston and May have pointed out, the ratio of taxonomists to the estimated number of species is 100 times higher for vertebrates than for invertebrates, and 10 times higher for the former when compared to vascular plants.[88] This results in many other curious situations. For instance, new bird species are discovered at the rate of about 3–5 per year, or about 0.03–0.05 percent of the total in the group; yet a tropical botanist can expect to find a new plant species for every 100 species collected. Collection of insect, fungi, marine macrofauna, and some other groups from previously unexplored habitats often results in the discovery that

[85] See Groombridge (1992).

[86] Recall that the definition of a "mass extinction" requires that this rate be higher than twice the background rate for many different groups.

[87] This discussion is based on the much more detailed one in May, Lawton, and Stork (1995).

[88] Gaston and May (1992). The numerical results reported are from this source.

Table 5.3.2. *Recent extinctions*

	Number of Species Certified Extinct since 1600	Number of Species Listed as Threatened	Recorded Extant Species (×1,000)	Percentage Extinct (%)	Percentage Threatened (%)
Animals					
Mollusks	191	354	100	0.2	0.4
Crustaceans	4	126	40	0.01	0.3
Insects	61	873	1000	0.006	0.09
Vertebrates	229	2212	47	0.5	5
Fishes	29	452	24	0.1	2
Amphibians	2	59	3	0.1	2
Reptiles	23	167	6	0.4	3
Birds	116	1029	9.5	1	11
Mammals	59	505	4.5	1	11
TOTAL	485	3565	1400	0.04	0.3
Plants					
Gymnosperms	2	242	0.8	0.3	30
Dicotyledons	120	17 474	190	0.06	9
Monocotyledons	462	44 21	52	0.9	9
Palms	4	925	2.8	0.1	33
TOTAL	584	22 137	240	0.2	9

Note: The table lists the number and percentages of species in major taxa that have either become extinct in recorded history (since 1600) or are threatened. "Threatened" includes the IUCN categories of "vulnerable," "endangered," and "probably extinct," but not "rare," "insufficiently known," or any other category of concern. (The numbers in the last three columns are necessarily approximate.)

Source: May, Lawton, and Stork (1995), p. 11.

from 20 percent up to 50–80 percent of the species are new to science. Of the sixty-one extinctions of insect species recorded in Table 5.3.2, thirty-three are of lepidopterans. This may show nothing more than the fact that butterflies are charismatic species.[89] The table routinely and seriously underestimates the number of extinctions in those taxa that have not traditionally been foci of attention. (b) Even for taxa that have traditionally been systematically studied, the documented extinctions recorded in Table 5.3.2 are probably severe underestimates. The problem with putative extinctions that have occurred in the last half-century has already been mentioned. The stringent criteria employed

[89] See May, Lawton, and Stork (1995), pp. 11–13, for a development of this argument.

by the IUCN and WCMC or under CITES prevent many extinctions from being recorded during the long period of purgatory. In spite of the so-called amphibian crisis (the increased pace of recent extinctions of amphibians),[90] Table 5.3.2 notes only two extinctions of amphibian species. Moreover, many species – even of birds and mammals – occur only in remote locations, and their extinction may go unnoticed. There are some species that are known from only a single collection site. (c) The geographical distribution of the data that are the source of Table 5.3.2 underscores how selective those data are and, therefore, how unrepresentative they may be. For instance, 67 percent of the recorded animal extinctions on continents are from Europe and North America, and 20 percent are from Australia; that leaves only 13 percent to be distributed over Africa, Asia, and South America, including their animal-rich tropical habitats. Roughly 67 percent of recorded continental plant extinctions are from North America and Australia; all forty-five extinctions from sub-Saharan Africa are from the Cape floristic region of South Africa. This means that Table 5.3.2 should not be interpreted to conclude that we are definitely not in the middle of a mass extinction or biodiversity crisis. Moreover, that table does not take into account the drastic changes to habitat that have been occurring since about 1945, changes that probably led to worries about such a crisis in the first place. To go any further will require an assessment of yet another extinction rate:

(iii) The *projected extinction rate* (again, for all species or for some selected taxa) will allow us to decide, along with an estimate of the proportion of the taxa lost, whether we are in the midst of a mass extinction or biodiversity crisis. Those who believe that we have a biodiversity crisis on our hands generally base that claim not on current extinction rates (or those from the recent past) but on projections of the rate of extinction in the near future. For instance, based on such a prediction, the Global 2000 Report to the President (of the United States) estimated that between 500,000 and 2,000,000 species could become extinct between 1980 and 2000.[91] If that estimate had been correct, there would have been little doubt that we were in the middle of a mass extinction. However, that estimate now seems irresponsibly alarmist: the number of extinctions that are known to have occurred are a tiny – though not accurately known – fraction of the estimate. Claims such

[90] See, for instance, Sarkar (1996).

[91] See Shaffer (1987), p. 69, who quotes the original source.

as that incorporated in the estimate of extinction just reported have the potential to destroy the credibility of biodiversity conservationists. Projected extinction rates are going to be no more reliable than the methods of extrapolation, and probably much less so, because the data typically will have even more sources of uncertainty than in the other cases.

There are three methods by which these estimates are made. (a) The most common method used to estimate future extinction rates is to estimate habitat loss and use the species-area curve. According to this method, habitat loss is first directly estimated. From this, the area of the remnant habitat is estimated. The species-area curve is then used to calculate the number of species that this area supposedly can support. This is compared to our estimates of the current number of species in order to calculate the percentage of loss. At the high end, estimates have suggested a 8–11 percent global species loss per decade; at the low end, a 1–5 percent loss.[92] Even the lower of these figures would strongly suggest that there is an ongoing mass extinction. The trouble is that the species-area curve is notoriously unreliable; therefore, there is little reason to trust these estimates.[93] (b) A second method is to use the IUCN/WCMC databases – not directly (because of the problems noted earlier) but indirectly, by comparing the lists in successive editions of the database. For instance, the number of animal species listed as "threatened" increased by 30 percent between 1986 and 1990, though only thirty-three such species satisfied the stringent criteria for being added to the list of extinct species.[94] This amounts to an extinction rate of about seven vertebrate species per year. At this relative rate,[95] if sustained, it would take only 7,000 years for the extinction of half of the known 4,500 vertebrate species. However, this approach is fraught with uncertainty on two counts: (1) as noted earlier, the data compiled in these databases were collected opportunistically and may not be representative, and (2) old extinctions are being added for the first time, and since the databases are less than fifty years old (which is about the time it takes for a species to get listed), the estimated rate is certainly an overestimate. In any case, the estimates obtained in this way are generally lower than those obtained from species-area curves

[92] See Groombridge (1992), p. 203, Table 16.3.
[93] Recall the discussion earlier in this chapter (in § 5.2).
[94] See Smith et al. (1993a, b) and May, Lawton, and Stork (1995) for details of this analysis.
[95] This is the ratio of the number of species becoming extinct to the total number of species present at that time.

but nevertheless high enough to suggest an ongoing mass extinction.
(c) For better-known categories, the data in the IUCN-WCMC database can be used in another way. The extinction probability distributions for these species as functions over time can be assessed using a variety of techniques.[96] From these probabilities, we can extrapolate tentatively to the general situation. The estimates obtained agree, more or less, with those in (b). However, this method is also fraught with the same uncertainties as those in (b).

It should be obvious that there is no clear answer to the question of whether there is an ongoing mass extinction (and therefore a "biodiversity crisis," if that is defined as a mass extinction). The current extinction rate, as calculated from the recorded extinctions, would deny that there is such a situation. But not only is that estimate fraught with uncertainly, it cannot take into account the increased destruction of habitat during the last fifty years. The projected extinction rates discussed in this subsection would all place the current era within such a period of mass extinction. But they are fraught with as much, if not more, uncertainty. Luckily, the case for biodiversity conservation that was constructed in Chapter 4 does not rely on there being such a crisis and, in fact, intentionally avoids alarmist talk.

96 See Mace (1994, 1995).

6

The Consensus View of Conservation Biology

Turning, finally, to the biological subdiscipline of conservation biology, in one sense it is possible not only to place and date, but even to time when that subdiscipline formally emerged as an organized field of research: at about 5 P.M. (EST), 8 May 1985, in Ann Arbor, Michigan, at the end of the Second Conference on Conservation Biology.[1] Two ad hoc committees, chaired by Jared Diamond and Peter Brussard, had met during the conference to discuss the need for a new society and a new journal. Following their reports, an informal motion to found the society was passed, and Michael E. Soulé was given the task of organizing it. A decision was also made to found a new journal, *Conservation Biology*. The fact that a successful European journal devoted to the same topic, *Biological Conservation*, had been in existence since 1968 apparently went unnoticed. In retrospect, participants at the Ann Arbor conference seem to have been strangely convinced that they were boldly going where no one had gone before. This reflected little more than the persistent myopia of the biodiversity conservation community in the United States during that period, a point that will be of importance later in the discussion of the relatively slow emergence of a consensus view of the aims and structure of conservation biology.

In North America, in order for conservation biology to emerge successfully as an academic discipline, things moved quickly from the Ann Arbor conference. In December 1985, Soulé published a long manifesto, "What Is Conservation Biology?," in *BioScience*, the journal most visible not only to academic but also, especially, to nonacademic biologists in the

[1] Soulé (1987); see also Sarkar (1998b) for a historical outline on which this chapter is partly based. There is also considerable overlap between this chapter and the more technical account given in Sarkar (2004).

United States. It defined the precepts of the movement for that broad audience and argued that a new interdisciplinary science, conservation biology, based on both substantive (that is, factual – Soulé called them "functional") and normative ethical foundations, had been created to conserve what still remained of our biological heritage. This science was ultimately prescriptive: it prescribed management plans for the conservation of biological diversity at every level of organization. Setting the tone for much of the discussion during the early years of North American conservation biology, Soulé emphasized that the new field was a "crisis discipline."[2] The first issue of the new journal, *Conservation Biology*, appeared in May 1987, and the first annual meeting of the new society was held at Montana State University in June 1987. Within ecology, Janzen's influential exhortation to tropical ecologists to undertake the political activism necessary for conservation, which was discussed in Chapter 2 (§ 2.1), fortuitously appeared in 1986.

Thus, between 1985 and 1987 conservation biology emerged in the United States as an organized academic discipline. Its focus soon became "biodiversity," a term that entered the everyday and scientific lexicons around 1988. This neologism was coined by Walter G. Rosen at some point during the organization of the 21–24 September 1986 National Forum on BioDiversity held in Washington, D.C., under the auspices of the U.S. National Academy of Sciences and the Smithsonian Institution.[3] The new term was initially intended as nothing more than a shorthand for "biological diversity" for use in internal paperwork during the organization of that forum. However, from its very birth it showed considerable promise of transcending its humble origins. By the time the proceedings of the forum were published, Rosen's neologism – though temporarily mutated as "BioDiversity"[4] – had eliminated all rivals to emerge as the title of the book.

The term "biodiversity" found immediate widespread use following its introduction. As Takacs has pointed out: "In 1988, *biodiversity* did not appear as a keyword in *Biological Abstracts*, and *biological diversity* appeared once. In 1993, *biodiversity* appeared seventy-two times, and *biological diversity* nineteen times."[5] The first journal with "biodiversity" in its title,

[2] Recall the discussion of "crisis" in Chapter 1 (§ 1.3) and Chapter 5 (§ 5.3). The claim that conservation biology is a "crisis discipline" is, surprisingly, uncritically accepted by Kingsland (2002) in her short discussion of the history of the emergence of today's consensus view. In most other aspects, Kingsland's reconstruction agrees with the one offered here.

[3] Historical material about "biodiversity" is taken from Takacs (1996).

[4] See Wilson (1988).

[5] Takacs (1996), p. 39; italics in the original.

Canadian Biodiversity, appeared in 1991; it changed its name to *Global Biodiversity* in 1993. A second journal, *Tropical Biodiversity*, began appearing in 1992; *Biodiversity Letters* followed in 1993. A sociologically synergistic interaction between the use of the term "biodiversity" and the growth of conservation biology as a discipline occurred, and it led to the reconfiguration of environmental studies that we see today, in which biodiversity conservation has emerged as a central focus of environmental concern. However, for all its appeal, "biodiversity" soon proved notoriously difficult to define, a point that will be discussed in detail in the last section of this chapter (§ 6.5).

In 1989, Soulé and Kohm published a primer on research priorities for the field. It was catholic in scope, including demography, ecology, genetics, island biogeography, public policy, and systematics as components of conservation biology. It called for massive biological surveys, especially in the neotropics, and for the circumvention of legal barriers to the use of U.S. federal funds for the purchase of land in other countries.[6] Soulé and Kohm viewed such acquisitions as critically important for biodiversity conservation in countries of the South. In 1993, Primack produced the first textbook of conservation biology; in 1994, Meffe and Carroll followed with their own, more comprehensive effort.[7] Their book began by repeating what, in the United States, had become a central dogma of conservation biology, that human population growth is *the* cause of the decline of biological diversity.[8] However, unlike many others at the forefront of the field in the North, they correctly mitigated this claim by explicitly noting that it is not human numbers per se that matter, but patterns of habitation and consumption. A sketch of the history of biological conservation with a rather narrow focus on the United States followed. Modern conservation biology – that is, the developments of the preceding decade – was declared to be a "synthesis" of all past attempts. A brave new science was born. Its brief was no less than to halt the decline of "biodiversity."

By and large, U.S.-style conservation biology viewed the field as emerging out of ecology, or at least out of ecology and some closely related academic disciplines, such as demography and population genetics. This is probably to be explained by the fact that most early conservation biologists in the United States, including all of those mentioned earlier, came

[6] Soulé and Kohm (1989).

[7] In retrospect, Primack (1993) appears to be a simplistic effort, whereas Meffe and Caroll (1994) is encyclopedic.

[8] Recall the discussion at the end of Chapter 1 (§ 1.3).

from academic or "pure" rather than applied scientific backgrounds.[9] What this approach largely missed is the fact that biodiversity conservation is as much a sociopolitical process as it is a scientific one. By and large, the most obvious mechanism of biodiversity depletion is habitat conversion. Habitat conversion occurs because, almost always, there are many potential uses for a piece of land (a particular place) other than its protection for biodiversity conservation. Many U.S. conservation biologists during this period tended to be purists, condemning all these other uses without a fair hearing.[10] What this attitude ignored is that the claims for other uses of land must be explicitly addressed in any sociopolitical context. Even ignoring ethical issues, at the purely prudential level, a refusal to enter the political process of negotiation and reconciliation is a recipe for disaster insofar as it invites habitat conversion by those who do enter the process but have little concern for biodiversity.

In sharp contrast to the United States, a radically different approach to biological conservation was pioneered during the same period in Australia by Graeme Caughley, Michael P. Austin, Christopher R. Margules, and Robert L. Pressey, among many others. In some ways, this approach reflected the unique Australian experience: relatively recent extensive habitat conversion, leaving much greater scope for successful biodiversity conservation compared to the situation in Asia, in Europe, or even in North America;[11] problems with the control of introduced species, and so on. More importantly, it reflected the practical background of many of these Australian researchers in the management of wildlife and other biological resources. This typically resulted in an explicitly pragmatic attitude toward the solution of biodiversity conservation problems. The role of academic ecology was limited in this context. Moreover, there was little explicit consideration of the ethically normative basis of conservation practice, again in sharp contrast to practice in the United States, even though, by paying attention to sociopolitical factors, this approach implicitly incorporated anthropocentric values at every stage of conservation planning. Until the late

[9] A proper sociological analysis remains to be done. Consequently, the claim made in the text must at present be regarded as conjectural, though guided by educated intuition. Being a "new" science, with most of its founders still alive, conservation biology provides truly novel opportunities for "meta"-scientists (historians, philosophers, sociologists, etc., of science). Strangely, few have availed themselves of these opportunities – see, however, Takacs (1996).

[10] Often, this reaction came hand in hand with an attribution of intrinsic value to biodiversity and a "resolute" biocentrism – recall the critical discussion of Chapter 3, § 3.4.

[11] See Margules (1989) for this argument as well as a detailed early accounting of Australian contributions to conservation biology.

1990s, reflecting the myopia mentioned at the beginning of this chapter, the Australian developments were largely ignored in the United States even while those methods were becoming increasingly popular in many other regions, including South Africa and the United Kingdom.

A 1989 volume of *Biological Conservation* edited by Margules brought the Australian approach to conservation biology to a broader audience.[12] Margules, Higgs, and Rafe's earlier (1982) criticism of the use of island biogeography theory for biological reserve network design was mentioned in Chapter 5 (§ 5.1); this criticism was reaffirmed in the 1989 volume.[13] The Australian school was also skeptical of the use of population viability analysis (PVA) to determine minimum viable populations (MVPs), which was central to U.S.-style conservation biology (and will be discussed later in this chapter, in section 6.4). In both cases, the Australians showed greater interest in what was empirically known about various models rather than in what was intuitively plausible about them – this was yet another salient difference between them and their counterparts in the United States. Systematic biodiversity reserve network design that went beyond the simple strategy of creating national parks became a central focus of the Australian framework. The emphasis was on the "efficient" selection of reserves (that is, the achievement of biodiversity representation goals using as little land as possible) and on the use of effectively measurable "surrogates" to estimate biodiversity.[14] The process began with the systematic collection of data through adequate and efficient sampling techniques.[15] The Australian school recognized that "surrogates" for biodiversity, parameters that can be readily and accurately assessed in the field, have to be chosen for rapid biodiversity assessment. Concern for the adequacy of surrogates also emerged early on, and the difficulties encountered in trying to solve this problem will be discussed later in this chapter (§ 6.3).

In the Australian framework, the design of networks of conservation areas was generally taken to occur in the presence of explicit constraints: for instance, a limit on the area that can be allocated to biodiversity conservation,

[12] *Biological Conservation* 50 (1989). Caughley and Gunn (1996) was the first textbook from the Australian perspective.

[13] Margules, Higgs, and Rafe (1982); Margules (1989).

[14] What was traditionally called "efficient" is better characterized as economical, to distinguish it from computational efficiency – see Sarkar et al. (2004). The term "surrogate" was apparently first used by Austin and Margules (1986). It was subsequently used in this way in 1989 by Mackey and colleagues (1989) in a discussion of the representativeness of the wet tropics in the Queensland World Heritage property.

[15] See Austin and Heyligers (1989) and Nicholls (1989); for a later paper emphasizing the importance of carrying out adequately designed surveys, see Haila and Margules (1996).

or a limit on the amount that can be spent on land acquisition for conservation. Consequently, places had to be prioritized for their biodiversity value. Conservation area networks could then be created by iteratively selecting places from the top of the prioritized list. As early as 1988, Margules, Nicholls, and Pressey published the first explicit (and efficient) algorithm for solving this problem.[16] It was based on a parameter called "complementarity," which will be discussed in some detail later in this chapter (§ 6.2). It called for the prioritization of places on the basis of what new – that is, unprotected – or inadequately protected species or other biodiversity surrogates a place would add to a conservation area network, rather than on richness (the absolute number of surrogates present in the place).

A 1994 piece by Caughley underscored the differences between the two traditions.[17] In particular, it systematized the Australian disquiet over PVA as practiced in the United States. Caughley argued that two "paradigms"[18] had emerged in conservation biology: a "small populations" paradigm and a "declining populations" paradigm. In PVA, the former dictated the use of stochastic models, the latter the use of deterministic models. Caughley argued that the former contributed little to the conservation of species in the wild because, beyond the trivial insight that small populations are subject to stochasticity, stochastic models do not help us to understand why species are at risk. Conservation had a better chance of success in the latter case, because with large populations it is usually possible to design field experiments with appropriate controls to determine the mechanisms of decline. Such experiments, in turn, can potentially lead to new and badly needed theoretical explorations within the declining populations paradigm.[19] Though Caughley did not couch his discussion in terms of national traditions, the small populations paradigm dominated the U.S. tradition in conservation biology, whereas the declining populations paradigm dominated the Australian tradition.

The period since 1995 has seen a new consensus framework for conservation biology emerging through the integration of insight from both traditions. In Australia, Margules and Pressey formulated this consensus view in 2000 in an article in *Nature*; in the United States, the Nature Conservancy presented a very similar framework in 2002 in *BioScience*, underscoring the

[16] Margules, Nicholls, and Pressey (1988).

[17] See Caughley (1994), which remains one of the methodological classics of conservation biology.

[18] As with most scientists, Caughley presumably borrowed this term from Kuhn (1962). It is put within quotation marks here to note the presence of many philosophical problems with its use. A discussion of these problems is beyond the scope of this book.

[19] For a response to Caughley, see Hedrick et al. (1996).

extent of the consensus achieved.[20] Central to this consensus is the idea that conservation biology is about systematic conservation planning through the "adaptive management"[21] of landscapes.[22] The actual framework that has been developed is one for the prioritization of places for their biodiversity *value*[23] and the formulation of management procedures for the long-term (in principle, infinite) survival of the biological units of interest. The entire process is supposed to be periodically iterated, because species (or other units) may have become extinct – or may have recovered from problems – in the interim, thereby changing the biodiversity value of a place, or because management practices may have turned out to be ineffective. This is the only sense in which the process is supposed to be adaptive. However, the framework is so new – it is yet to be fully implemented anywhere – that we have no idea whether this requirement of being adaptive will necessitate any change in the framework as it is currently understood. At present, "adaptive management" is only a slogan embodying a tantalizing promissory note.

6.1. ADAPTIVE MANAGEMENT

As noted earlier, the new consensus framework for adaptive management was first clearly articulated by Margules and Pressey; Table 6.1.1 expands their framework into a ten-stage one in order to remove some lacunae.[24] It defines the practical strategies and aims of conservation biology. The basic framework is one for selecting a network of places for conservation

[20] See Margules and Pressey (2000), which will be analyzed in detail in § 6.1; Groves et al. (2002) presented the Nature Conservancy's framework. Unfortunately, some of the terminology in Groves et al. (2002) is out of date. For instance, "target" is used for "surrogate." However, unlike Margules and Pressey (2000), it draws explicit attention to the viability problem, for which it should be recommended. Slightly earlier, and independent of Margules and Pressey, Craig L. Shafer (1999) of the United States National Park Service provided a sketch of a similar framework for selecting national parks, primarily but not entirely in the context of the United States. However, Shafer's outline was not as systematic as that of Margules and Pressey (2000). In passing, it should be noted that the consensus is not complete – see Soulé and Terborgh (1999) for the rather different approach of the "Wildlands Project" for North America.

[21] The term "adaptive management" was invented by Holling (personal communication) in 1975. See also Holling (1978) for its first appearance in print.

[22] In what follows, it will simply be assumed that the appropriate focus of biodiversity conservation is the use of the surface of Earth (land or oceans). Wood (2000, Chapter 2) argues for this point in more detail.

[23] This is to be distinguished from prioritization for biodiversity *content* (see § 6.1).

[24] Margules and Pressey (2000), p. 245, present a six-stage framework, Groves and colleagues (2002) a seven-stage one. Shafer (1999) lists nine points, which, however, cannot be interpreted as a sequential set of stages.

action on a "landscape scale" and then devising management strategies. (Here, "landscape scale" is being interpreted loosely to refer to spatially large habitats that will contain several distinct populations of species with large body sizes.) Traditionally, in the United States and many other areas of the North, the action to be taken was presumably the exclusion of human habitation and extractive exploitation, exemplified by the creation of national parks. What is perhaps most important about the new consensus framework is the extent to which the range of options has expanded, both with respect to the designation of places on the basis of their relevance to biodiversity and with respect to the type of management practices that are considered. Instead of national parks and reserve networks, attention shifts to "conservation area" networks, with a conservation area defined as any place at which some conservation action is implemented.[25]

Note that reservation – for instance, in national parks – for biodiversity is just one of many possible conservation actions. It may not be possible in many regions and may not even be desirable. If the biodiversity of a place can persist in the presence of certain forms of agricultural use – which is often the case, for example, with some forms of bird biodiversity – then it is important that the place not be converted into a built landscape; but it is irrelevant, and potentially even dangerous (recall the discussion of Keoladeo National Park in India in Chapter 2, § 2.3) for it to be converted into a national park requiring human exclusion. In many such situations, the designation of a national park will be a particularly inefficient and inappropriate use of limited resources.

The consensus framework will be analyzed in this section: it includes both theoretical and empirical components. Each stage of the framework involves a particular set of actions to be performed, though as will be clear from the discussion of surrogacy (§ 6.3), these stages should not be viewed as necessarily sequential. The structure of the theoretical components (Stages 2 and 5–8 of that framework) will be briefly elaborated in this section; their further analysis will occupy the rest of this chapter and part of the next.[26] The emphasis will be on the role played by conventions in the construction of the consensus framework: choices will have to be made on pragmatic grounds, though not arbitrarily (and certainly not subjectively, if "subjective" is taken to refer to mere unelaborated personal

[25] Sarkar (2003) makes the case for this shift in terminology.

[26] The discussion will largely follow that of Sarkar (2002) and Sarkar (2004). Stage 1 here corresponds to Stage 1 of Margules and Pressey (2000) and to Step 2 of Groves et al. (2002); it is also point 2.5 of Shafer (1999, p. 125).

Table 6.1.1. ***Systematic conservation planning***

1. **Compile and assess biodiversity data for region:**
   - Compile available geographical distribution data on as many biotic and environmental parameters as possible at every level of organization;
   - Collect obviously relevant new data to the extent feasible within available time; remote sensing data should be easily accessible; systematic surveys at the level of species (or lower levels) will usually be impossible;
   - Assess conservation status for biotic entities, for instance, their rarity, endemism, and endangerment;
   - Assess the reliability of the data, formally and informally; in particular, assess the process of data selection.
2. **Identify biodiversity surrogates for region:**
   - Choose true surrogate sets for biodiversity for part of the region; be explicit about criteria used for this choice;
   - Choose alternate sets of estimator-surrogates that can be (i) measured (quantified) and (ii) easily assessed in the field (using insights from Stage 1);
   - Prioritize places using true surrogate sets;
   - Prioritize places using as many combinations of estimator-surrogate sets as feasible;
   - Assess which estimator-surrogate set is best on the basis of (i) efficiency and (ii) accuracy.
3. **Establish conservation targets and design goals:**
   - Set quantitative targets for surrogate coverage;
   - Set quantitative targets for total network area;
   - Set quantitative targets for minimum size for population, unit area, etc.;
   - Set design criteria such as shape, connectivity, and dispersion;
   - Set precise goals for criteria other than biodiversity.
4. **Review existing conservation areas:**
   - Estimate the extent to which conservation targets and goals are met by the existing set of conservation areas.
5. **Prioritize new places for potential conservation action:**
   - Prioritize places for their biodiversity content in order to create a set of potential conservation area networks;
   - Optionally, starting with the existing conservation area networks as a constraint, repeat the process of prioritization in order to compare results;
   - Incorporate design criteria such as shape, size, dispersion, and connectivity.
6. **Assess prognosis for biodiversity for each potential targeted place:**
   - Perform population viability analysis for as many species using as many models as feasible;
   - Perform the best feasible habitat-based viability analysis in order to obtain a general assessment of the prognosis for all species in a potential conservation area;
   - Assess vulnerability of a potential conservation area to external threats, using techniques such as risk analysis.
7. **Refine networks of places targeted for conservation action:**
   - Delete the presence of surrogates from potential conservation areas if the viability of that surrogate is not sufficiently high;

Table 6.1.1 (*continued*)

- Run the prioritization program again to prioritize potential conservation areas by biodiversity value;
- Incorporate design criteria such as shape, size, dispersion, and connectivity.

8. **Perform feasibility analysis using multiple criterion synchronization (MCS):**
   - Order each set of potential conservation areas by each of the criteria other than biodiversity;
   - Find all best solutions;
   - Discard all other solutions;
   - Select one of the best solutions.
9. **Implement conservation plan:**
   - Decide on most appropriate legal mode of protection for each targeted place;
   - Decide on most appropriate mode of management for persistence of each targeted surrogate;
   - If implementation is impossible, return to Stage 7;
   - Decide on a time frame for implementation, depending on available resources.
10. **Periodically reassess the network:**
    - Set management goals in an appropriate time frame for each protected area;
    - Decide on indicators that will show whether goals are met;
    - Periodically measure these indicators;
    - Return to Stage 1.

preference).[27] If the consensus framework survives for any significant length of time, these conventional choices will eventually appear to be so natural that their genesis in choices based on pragmatic grounds will probably be forgotten. Such recourse to convention will no doubt be anathema to scientific realists and others who believe that scientific concepts must have a "one-to-one correspondence" to the furniture of the world (independent of science). However, conventional elements are part of every science, among other reasons because theories are underdetermined by experimental evidence. Conventions become reified as they become entrenched through the instrumental success of theories. Conservation biology is not special in this respect; for philosophers of science, what is interesting is that they can watch the explicit process by which conventional elements get reified as a new science gropes for maturity.

The stages of systematic conservation planning are presented in Table 6.1.1. What is most striking about this framework is how different it appears from the traditional ecology that was discussed in Chapter 5 (especially in § 5.1). While Stage 1 seems obvious, data surveys have often

[27] Conventionality does not imply arbitrariness, a point that is often lost and cannot be overemphasized.

been inadequately designed.[28] In particular, as the discussion of place prioritization procedures will show (in § 6.2), reports of raw data from surveys must retain lists of surrogates from surveyed places. For example, data treatment such as retaining only summary statistics, such as the surrogate richness of a place, is insufficient. Stage 2 concerns the choice of appropriate surrogates to be used to represent biodiversity in the planning process – as will be emphasized later (in § 6.3), this choice raises interesting and difficult theoretical and empirical questions.[29]

Stage 3, as Table 6.1.1 notes, involves some conventional choices about targets to be set.[30] Without explicit targets and goals, it would be impossible to assess the success (or failure) of a conservation plan. However, the choice of such targets and goals is not entirely determined by biological considerations and provides ample scope for controversy. For instance, two types of targets are typically, used: (i) a level of representation for the expected coverage of each surrogate within a conservation area network, and (ii) these representation targets along with the maximum area of land that can be put under a conservation plan.[31] However, the actual numbers used are not determined – or even strongly suggested – by any biological criteria (models or empirical data). Rather, they represent conventions arrived at by educated intuition: type (i) targets are supposed to reflect biological knowledge, while type (ii) targets are usually the result of a "budget" of land that can be designated for conservation action. Soulé and Sanjayan have recently forcefully argued that these targets may lead to an unwarranted sense of security, because their satisfaction can potentially be interpreted as suggesting that biodiversity is being adequately protected when in fact the targets may reflect only sociopolitical constraints, such as a budget.[32] The issue remains controversial.

[28] Haila and Margules (1996) emphasize this surprising point in the general context of ecology.

[29] This stage is not explicitly distinguished by Margules and Pressey (2000), Groves et al. (2002), or Shafer (1999).

[30] This is Stage 2 of Margules and Pressey (2000), Steps 1 and 3 of Groves et al. (2002) – it is unclear why these steps are separated; it is also point 2.1 of Shafer (1999, p. 124). Margules and Pressey call it "subjective," which is unfortunate, since these conventional choices will usually involve public discussion of the adequacy of the set targets and will be guided by past experience and ecological knowledge.

[31] A common target of type (i) is to set the level of representation at 10 percent for the coverage of each surrogate – see, for instance, IUCN (1983), which does not offer a biological basis for this target. A common target of type (ii) is 10 percent of the total area of a region – see, for instance, Araújo et al. (2001).

[32] However, Soulé and Sanjayan (1998), though presenting a valid problem, probably overstate their case – as will be indicated in section 6.3, considerable biological knowledge can enter the process of specifying targets.

Stage 4 is simply a review to see which targets and other design goals from Stage 3 are already met in the existing network of conservation areas.[33] If such a network is being created for the first time for a region, then this stage is irrelevant; however, the typical situation is one in which the task is that of extending and refining an existing set of conservation areas. If, in the process of reviewing existing protected areas, it is found that some protected places contribute little to the conservation of biodiversity, a rational policy choice dictates that these places be targeted for potential "delisting," that is, for removal from the set of places that merit conservation action. This is particularly important because, in most regions, national parks and other such reserves have traditionally been designated with no explicit concern for biodiversity in mind. Rather, they have often been selected to protect sublime landscapes or for their recreational possibilities, or sometimes simply because the land had no other potential use. This is the process of inefficient ad hoc reservation that provided the impetus for the formulation and promotion of systematic, algorithmic conservation area network design, which is one of the main virtues of the new consensus framework.[34] As an example of inefficient design, consider Figure 6.1.1, a GIS-generated image that shows the national parks of Québec and the distribution of faunal and floral species considered to be at risk.[35] Except for the national parks in the Gaspé peninsula, there is almost no record of species at risk within the national parks and reserves; these conservation areas are, therefore, of low priority from the perspective of biodiversity conservation. If there is a constraint on how much land can be targeted for conservation action, what Figure 6.1.1 suggests is that some of Québec's current national parks should perhaps be replaced by places outside the park system that have higher biodiversity value.

Stages 5–8, along with Stage 2, form the theoretical core of conservation biology and will be further discussed below. Stage 9 moves from theory to practice, trying to set the form of management responses to what was found at the end of Stage 8.[36] Finally, Stage 10 emphasizes that adaptive management is an iterative process, requiring the monitoring and adaptation

[33] This is Stage 3 of Margules and Pressey (2000) and Step 4 of Groves et al. (2002); it is not explicitly mentioned by Shafer (1999).

[34] Pressey (1994) argues forcefully against ad hoc reservation. Prendergast, Quinn, and Lawton (1999) have recently questioned the value of systematic conservation area network design, arguing that the money spent in gathering the necessary data and performing the analyses is better spent acquiring land on the basis of expert opinion; for a response, see Pressey and Cowling (2000), who reinforce the arguments of Pressey (1994).

[35] This example is based on Sarakinos et al. (2001).

[36] This is Stage 5 of Margules and Pressey (2000); it is surprisingly ignored by Groves et al. (2002) and Shafer (1999).

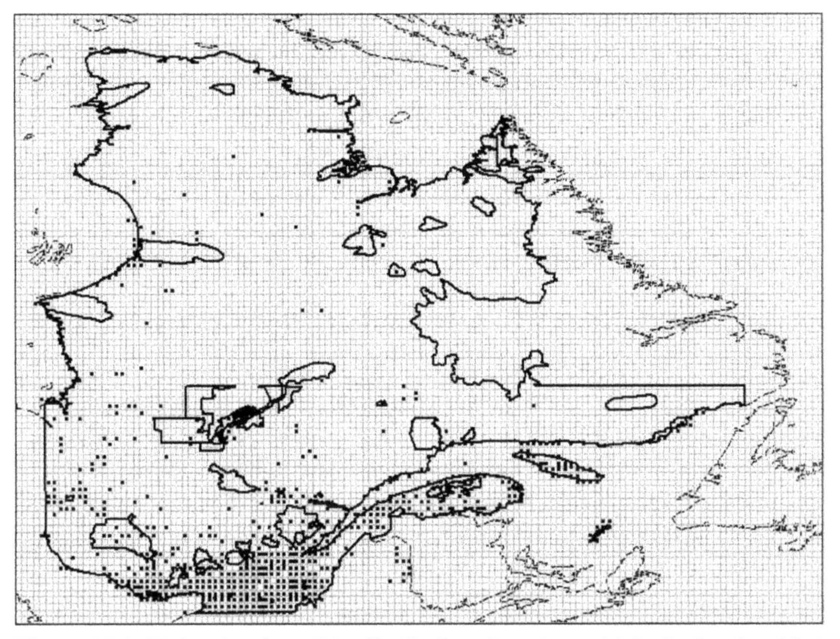

Figure 6.1.1. National parks and the distribution of species at risk in Québec (modified from Sarakinos et al. [2001]). From the perspective of biodiversity conservation, the present network of national parks and reserves in Québec, the boundaries of which are shown in black, is inadequate because almost all records of species (fauna and flora) at risk, shown as black points, are from areas outside the network and, except in the Gaspé peninsula (to the south central), there are very few records inside them. (The data points are presence/absence and, therefore, reliable.)

of management policy depending on the responses of management units to current practices.[37]

Stages 2 and 5–8 have become the most systematically developed theoretical part of conservation biology in this framework. They involve the solution of four problems that will be analyzed in the rest of this chapter and the next:

(i) Stages 5 and 7 concern the *place prioritization* problem:[38] places must be prioritized on the basis of their biodiversity *content* so that this prioritized list, after further analysis, can be used as the basis for the

[37] This is Stage 6 of Margules and Pressey (2000); again, it is surprisingly ignored by Groves et al. (2002) and Shafer (1999), leaving those formulations open to the charge that they are not ultimately about adaptive management.

[38] Together, they comprise Stage 4 of Margules and Pressey (2000); they correspond to Steps 6 and 7 of Groves et al. (2002) and to points 2.6 and 2.7 of Shafer (1999, pp. 126–127).

selection of places to include in a network of conservation areas, that is, places destined for conservation action of some sort. Place prioritization is critical to biodiversity conservation because, due to limited resources, not all places that are of some biological interest can be conserved in practice. More theoretical work has been done on place prioritization than on any other area within conservation biology; it will be discussed in detail in section 6.2. How place prioritization enters into the formulation of an operational definition of "biodiversity" will be discussed in section 6.5.

(ii) Solving the *surrogacy* problem forms Stage 2 of the framework presented here: features must be selected by which biodiversity may be estimated; such a feature is called a biodiversity "surrogate." Potential candidates include the set of species or other biotic entities, the patterns of complexity of trophic webs, and so on. The surrogacy problem is nontrivial and will occupy section 6.3. It will be argued there that place prioritization must be used as part of the solution to the surrogacy problem. This is why it is being listed here after the place prioritization problem.

(iii) Stage 6 consists of solving the *viability* problem:[39] once there is a list of places prioritized on the basis of biodiversity content, we can proceed to assess the long-term prognosis for the biological entities of interest in them, for instance, populations of species.[40] The biodiversity *value* of a place depends on its rank by biodiversity content and this assessment. Once the biodiversity values of places are known, a new place prioritization can be established using them. The viability problem is as difficult as it is important. It will occupy section 6.4.

(iv) Stage 8 consists of solving the *feasibility* problem:[41] to return to a theme that runs throughout this book, biodiversity is not the only criterion that

[39] This stage is ignored in Margules and Pressey's (2000) formulation, where there is no explicit emphasis on the problem of devising appropriate conservation policies for implementation; it is partially subsumed under their Stage 4. The main limitation of that formulation is that too many heterogeneous steps, corresponding to Stages 5–8 of the framework here, are included as a single stage (Stage 4). However, Margules and Pressey assume that viability considerations remain part of the common background of the entire planning process. For Groves and colleagues (2002), this is Step 5. The same process is spread over points 2.4, 2.8, and 2.9 of Shafer (1999, pp. 124–125, 127–128).

[40] Logically, we can attempt to solve the viability problem without place prioritization. However, in practice we would be unlikely to do so. Unless we are thinking of protecting some places and not others, there is little reason to attempt something as complicated as viability analysis.

[41] It is not given explicit recognition by Margules and Pressey (2000) but rather, as noted above, subsumed under their Stage 4; it is ignored by Groves et al. (2002) and Shafer (1999).

matters when the fate of a place must be decided. Sociopolitical (including economic) and other criteria, sometimes including wilderness status, also matter. Using the terminology introduced in Chapter 4 (§ 4.1), not only are these criteria often *incommensurable* in the sense that they cannot all be reduced to the same (unitary) scale, they may also be *incompatible* in the sense that the optimization of one necessarily involves a nonoptimal change in another. Establishing feasibility involves attempting to synchronize these multiple criteria to the extent that this is possible. If they cannot be so synchronized to an acceptable extent, then no adequate conservation policy can be devised. At that stage, all values and rankings must be reexamined and the entire process reiterated. Very little fully satisfactory work has been done on multiple criterion synchronization in the context of biodiversity conservation; technical aspects of this problem will occupy the first half of the next chapter (§§ 7.1–7.3). However, even if all these technical problems are successfully solved, section 7.1 will emphasize that not all criteria may be amenable to synchronization because some may be *intangible* (in the sense of Chapter 4, § 4.1).

From a philosophical perspective, as noted before, what is perhaps most striking about this framework is the extent to which it departs from traditional scientific ecology by interweaving sociopolitical, often pragmatic, considerations with biological ones. With some rhetorical excess, we would be justified in arguing that conservation biology is as much a social science as a biological science.

6.2. THE PLACE PRIORITIZATION PROBLEM[42]

Place prioritization is critical to biodiversity conservation because not all places that are of some biological interest can be conserved in practice. Talk of "place" takes us into intuitively trivial but, strangely, relatively uncharted territory philosophically. A place is geographically rooted, and it loses its sense of place as it is generalized about. But places – precise biogeographical locations with their specific components, including all biological and cultural features – are what matter for conservation: they alone retain the heterogeneity that provides the intuition for bio*diversity*. A preference for

[42] For a more detailed account of the mathematical issues connected with place prioritization, see Sarkar et al. (2004).

a place is not merely a preference for an ecosystem or even a habitat,[43] both of which are supposed to admit abstract characterization: the same habitat at different places may hold a different complement of genes, species, communities, or whatever other unit that may be of conservation interest. Worry about unique entities such as places takes us in a direction opposite to that of conventional scientific generalization. Philosophy – historically, and, at least, the contemporary analytic tradition in Western philosophy[44] – has largely followed science in the deification of generality. Consequently, places have seemed inappropriate as loci of philosophical or scientific interest. Nevertheless, in the context of biodiversity conservation, we have to worry about the peculiarities of individual places – what entities they contain, what processes they admit, and what constrains those processes, all of which is subject to the contingencies of biological and geophysical history. As Aldo Leopold put it in his inimitable way: "One cannot study the physiology of Montana in the Amazon."[45] Since we have neither the economic nor the human resources to conserve every place of any biological interest – we would end up wanting to conserve Earth itself – plans for conservation must ultimately involve a prioritization of places.

Place prioritization begins with a complete list of biodiversity surrogates. It will be clear from the following discussion that summary statistics such as richness are insufficient. The critical consideration is that of economy (usually called "efficiency" in the biological literature – recall the discussion of § 6.1): the representation of as many different and as a high a concentration of individual biodiversity surrogates as possible in the least number of places. The motivation for such economy is that there is a limit to the number of places that can be targeted for conservation action. Generally, this assumption is made because all places come with a cost. In the procedure described later in this section, where all places are treated on a par, it is also implicitly assumed that all places come with the same cost, which is not a reasonable assumption in most circumstances. An extension of this procedure that does not make this assumption will be discussed in the next chapter (Chapter 7, § 7.3), and many other alternatives are possible.

[43] Both "ecosystem" and "habitat" are unusually vague terms; "ecosystem" used to be generally understood in terms of an ecosystem ecology that used physical variables (most notably, energy flows) to describe spatially extended biological systems (recall the discussion of Chapter 5, § 5.2). In that context, "habitat" was more rooted in place, but habitats are also classified by type (wetland, tropical wet forest, and so on). These days the two terms are often used interchangeably.

[44] See, in contrast, Malpas (1999). In general, the so-called continental tradition in Western philosophy has paid much more attention to "place" than the analytic one.

[45] See Leopold (1949), p. 196.

For systematic place prioritization, explicit targets must be set for the representation of each surrogate in a set of places designated for conservation action. As noted earlier, though these targets are in part conventional, they usually reflect considerable biological knowledge. For instance, if species are used as surrogates, a target of zero may be appropriate for very common or introduced species.[46] For endangered, fragile, or rare species, protecting every population may be the appropriate target. Given a list of places, a list of surrogates for each place, and a target for each surrogate, the place prioritization problem takes two canonical forms (which, for ease of expression, will be called the two "canonical problems"):

I. Given a set of places Σ, a set of surrogates Λ, and a list of surrogates $\lambda_i \subseteq \Lambda$ for each place $\sigma_i \in \Sigma$, and a target τ_i for each $\lambda_i \subseteq \Lambda$, prioritize all places in Σ in such a way that the targets are all met in a set Π of the lowest cardinality that is possible.[47]
II. Given a set of places Σ, a set of surrogates Λ, and a list of surrogates $\lambda_i \subseteq \Lambda$ for each place $\sigma_i \in \Sigma$, and a target τ_i for each $\lambda_i \in \Lambda$, and an integer k, find the set Π of cardinality k which maximizes the number of surrogates for which the target is met.

In the first canonical problem, economy is at stake, but there is no explicit constraint on the total number of places that can be selected. In the second canonical problem, an explicit constraint in the form of an upper limit to the number of places that can be conserved is present. (There are many variants to these two forms; they will not be considered here because they do not raise any new philosophically interesting issues.) If the targets are set higher than the number of places at which each surrogate occurs in the data set, all places in Σ will be ordered by priority. (Both canonical forms will return the same answer.) In section 6.5 it will be argued that this procedure implicitly operationally defines "biodiversity."

If all targets are set equal to one, the first canonical problem is known in computer science as the "location set covering problem"; the second is known as the "maximal covering location problem."[48] Exact algorithms

[46] For instance, Sarakinos and colleagues (2001) set a target of zero for the domestic mouse, *Mus musculus*, in their formulation of a conservation priority plan for Québec. An introduced species, facing no threat, *Mus musculus* deserved no attention compared to other species.

[47] For simplicity, it is being assumed that all places have the same area; taking account of differing areas requires only a minor modification of the procedure having no philosophical ramifications.

[48] See Church, Stoms, and Davis (1996) and ReVelle, Williams, and Boland (2002). Because of the relevance of such combinatorial optimization problems in the design of place prioritization procedures, Kingsland (2002) argues that techniques from operations research are being

(that is, ones that are guaranteed to produce the optimal or best solution) using integer programming techniques exist for both problems.[49] The trouble is that these algorithms are slow; the problem is known to be NP-complete.[50] A typical data set of about 10,000 places and 1,000 surrogates often takes an hour or more to resolve, even with the fastest desktop computers now available.[51] Since place prioritization characteristically involves the formulation of dozens of alternative prioritized sets of places potentially designated for conservation action (see the next chapter), the high computational times required militate against the use of these exact algorithms.

Starting with a pioneering algorithm published by Margules, Nicholls, and Pressey in 1988, the last two decades have seen a large number of heuristic algorithms designed to solve the place prioritization problem efficiently (that is, using as little computational time as possible).[52] These algorithms have primarily been based on three rules or principles for place prioritization:

(i) *Rarity*:[53] the rarity of a surrogate is the inverse of the frequency with which it occurs at different places in a data set. The rarity value of a place is determined iteratively by the rarity of the surrogates in it;

co-opted in conservation biology. This is a contentious interpretation of these developments on two grounds. (i) At most, only the exact algorithm has such a pedigree. The much more popular and useful heuristic algorithms discussed next in the text have their origin in a purely biological context. And (ii) the study of combinatorial optimization problems was neither initiated in, nor historically restricted to, operations research. It has always been part of theoretical computer science and has deep roots in combinatorial mathematics.

[49] Cocks and Baird (1989) were the first to point this out, after the heuristic algorithms discussed next in the text had been devised; see also Underhill (1994). Arthur and colleagues (1997) have produced a method for finding all optimal solutions of the first canonical problem when they are degenerate. There continues to be some controversy over the value of these exact algorithms – see Sarkar et al. (2004) for a discussion.

[50] An algorithm is NP (*n*on-*p*olynomial) if the time taken for execution grows faster than any polynomial function of the size of the input. (The size of the input is the size of the problem to be solved, for instance, whether 10 or 10,000 places have to be prioritized.) Roughly, a problem is NP-complete not only if, is there no known polynomial time algorithm to solve it but, moreover, if one were to be found, then all NP-complete problems could be solved in polynomial time. See Garey and Johnson (1979).

[51] For an extensive analysis of the computational performance of these algorithms, see Sarkar et al. (2004).

[52] Margules, Nicholls, and Pressey (1988). One interesting family of heuristic algorithms, based on simulated annealing, will not be discussed here because of lack of space. See, Possingham, Ball, and Andelman (2000).

[53] This is "rarity" defined using geographical range (recall the discussion of Chapter 3, § 3.2). The other definitions of "rarity" have so far not been implemented in place prioritization algorithms but merit further exploration.

the one with the rarest surrogate is ranked highest, with ties[54] being broken using the next rarest surrogate iteratively.

(ii) *Complementarity*: the complementarity rank of a place is the number of surrogates in it that have not yet met their targets in the currently designated places in the priority list.

(iii) *Richness*: the richness of a place is simply the number of surrogates in it.

Both richness and rarity can be captured by a summary statistic: richness is a number; rarity, as defined here, assigns each place a rarity rank. It is complementarity that requires the entire list of surrogates at each place. It should be clear that the iterative use of richness may not lead to the most efficient choice of places: two places may have very high richness but may have very similar surrogate compositions. Adding both to a set of designated places may yield much less total surrogate coverage than adding one of them to another with low richness but a highly different surrogate composition, that is, high complementarity. The principle of complementarity is critical to economical surrogate coverage.[55] That complementarity should lead to economy should come as no surprise. More surprisingly, it has turned out to be the case – shown by comparative studies by many groups on many data sets – that rarity also leads to efficient place designation. It remains unclear whether some important ecological insight looms behind this result.

A place prioritization algorithm usually consists of two steps: (i) an initialization step, during which the potential designated set is "seeded" with an initial place or set of places, and (ii) an iterative step, during which a single place is added to it. In practice, the initialization step will often involve inclusion of an existing conservation area network, as explicitly noted in the consensus framework discussed in section 6.1. If this is not the strategy that is used, then the use of complementarity suggests that richness should be used for initialization, since complementarity reduces to richness when there is as yet no place in the designated set. However, as an empirical matter, it has been found that rarity performs slightly better with

[54] There is a "tie" when more than one place performs equally well under the criteria so far deployed, for instance, when each of them has the rarest surrogate.

[55] It is hardly surprising, therefore, that this principle was independently formulated at least four times during the same decade: by Kirkpatrick (1983) in Australia; by Ackery and Vane-Wright (1984) in the United Kingdom; by Margules, Pressey, and Nicholls (1988), again in Australia; and by Rebelo and Siegfried (1990) in South Africa. The term "complementarity" was introduced by Vane-Wright, Humphries, and Williams (1991). For a history of the use of complementarity in systematic biodiversity planning, see Justus and Sarkar (2002).

respect to economy than does richness. The iterative step involves the use of rarity and complementarity (with differing precedence in different versions of this algorithm); richness is never used. For the first canonical problem, the iterative step is terminated when all surrogates have met their targets (or when no places remain to be prioritized).[56] For the second canonical problem, it is terminated when the maximum number of places is reached.

Sometimes, at both the initialization and iterative steps, a tie remains after rarity and complementarity have been taken into account. This situation allows the incorporation of more biological and other criteria. A typical biological criterion is adjacency: preference is given to a place that is adjacent to some place already included in the selected set. This would lead to bigger conservation areas and makes sense if the places being prioritized generally have small areas. A typical nonbiological criterion is economic cost, with a place with lower cost getting preference over one with higher cost. There are many other options.

In addition to computational speed, these heuristic algorithms have another advantage over exact algorithms: transparency. The heuristic procedure makes it clear why a place is selected, that is, what rule privileges it over other places during the iterative prioritization process. The exact algorithms mentioned earlier cannot produce this information, since they do not use rules that have any direct biological interpretation. Thus, when such heuristic algorithms are used, should there be several potential sets of designated places, further analysis, perhaps using expert opinion, can analyze whether the rules in the heuristic algorithms were being used appropriately in the given context, for instance, whether the adjacency rule was being used too often in a particular solution. If so, experts can choose to reject that plan. The point that deserves emphasis is that place prioritization procedures are not supposed to be used mechanically; some undeserved criticism has been directed against them on this ground.[57] Various place prioritization procedures are supposed to provide options for further analysis – for instance, the use

[56] Alternatively, in some implementations of this algorithm – for instance, ResNet (Aggarwal et al. 2000) – the iterative step can be terminated if a maximum cost or area is exceeded. These implementations incorporate considerations other than biodiversity into the place prioritization procedure. While, in a sense, this violates a certain kind of academic purism, it allows partial incorporation of other values that will be explicitly discussed in the next chapter. (If all places have the same area, termination by maximum area constitutes a solution to the second canonical problem.)

[57] See the criticism of Prendergast, Quinn, and Lawton (1999). For an appropriate response, see Pressey and Cowling (2000).

of other criteria than biodiversity content (or even value), which will be discussed in the next chapter.

Place prioritization procedures have also been criticized on two other grounds.[58] (i) They require data that are impossible to obtain. Whether this is so depends on the solution of the surrogacy problem, which will be discussed in section 6.3. (ii) The data they use are usually unreliable and should not form the basis for such analyses, which have serious practical consequences. The second criticism reflects a serious problem. The surrogate data for each place is supposed to be the record of a presence or absence of that surrogate. If the data used are modeled data – that is, the output of models that attempt to predict the distribution of the biodiversity surrogates – then such data are no more reliable than the models that produced them. In general, most planners correctly prefer to use observed data over modeled data. However, if the observed surrogates are environmental features (see the next section), then their reliability is no better than the reliability of the techniques used to gather that data. But there is worse to come. If the surrogates used are species or other taxa, then these "presence-absence" data should be the result of detailed systematic surveys. Except for a very few groups – most notably birds and butterflies, for which hobbyists inadvertently generate a large amount of usually reliable data – such data are not available from most places in the world outside the North. For many species, all that are available are modeled distributions, which are, once again, no more reliable than the models themselves. Usually all that are available are records from museums and casual sightings, "presence-only" data. Using such data with presence-absence data, or even by themselves, is fraught with uncertainty. Presence-only data are biased in a wide variety of ways, one of which, ease of access, will be illustrated later. At present there is no systematic way to counter the effects of such bias.[59] At the very least, place prioritization procedures must cope with this uncertainty. So far little work has been done toward that end. The next chapter (§ 7.5) will include one potential way

[58] The usual alternative is to use local expert opinion. However, nothing prevents the use of such opinion along with these systematic procedures, as envisioned in the previous paragraph. Cabeza and Moilanen (2001) review these objections.

[59] However, if all that is required is adequate representation of the biodiversity surrogates in a conservation area network, the use of presence-only data does not call into question the success of representation. Only places with surrogates present but unrecorded would be excluded from consideration. Thus, any solution using presence-only data will ensure representation but may not be an economical solution, because possibly better places were excluded from the analysis – see Sarakinos et al. (2001) for more details of this argument.

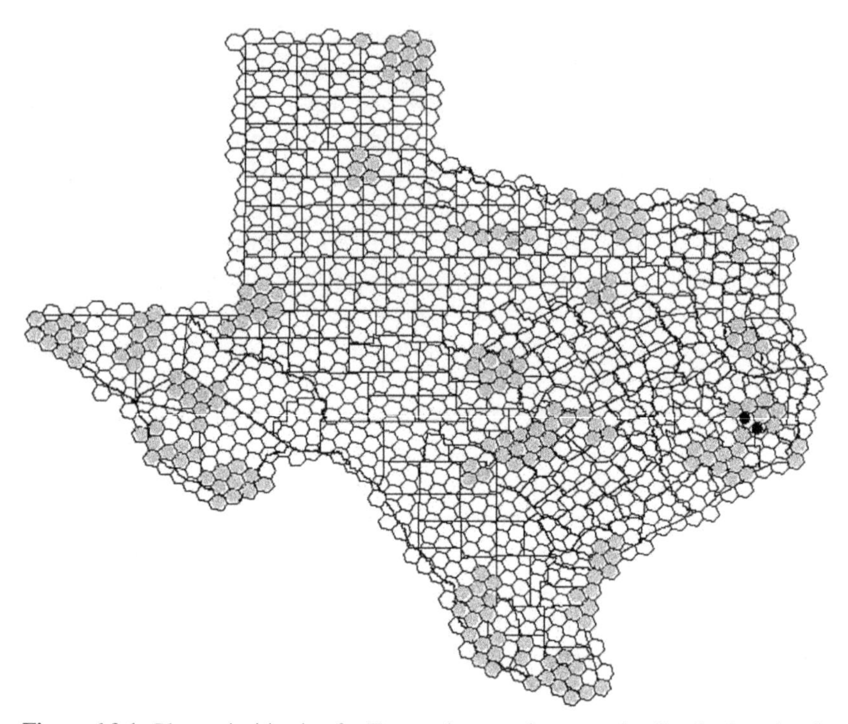

Figure 6.2.1. Place prioritization for Texas using vertebrate species distributions (modified from Sarkar et al. [2000]). The data set is discussed in the text. A target of representation of ten was set for each vertebrate. The prioritization used rarity first and complementarity second during the iterative step. Initialization was by rarity. Finally, a redundancy check was done at the end to see if any previously designated cell had become redundant through the inclusion of new cells, that is, if it could be dropped without any surrogate failing to meet its target. Redundant cells are shown in black. The designated cells are shown in grey.

to incorporate uncertainties, as reflected by probabilistic expectations, into these procedures.

Two examples will conclude this section. Figure 6.2.1 shows the results of a place prioritization for Texas using data from the Texas GAP Analysis Project.[60] Modeled distributions of 655 vertebrate species have been used as surrogates. Since these are modeled data, they can be treated as presence-absence. However, the results of these models are necessarily probabilistic, a feature that was not taken into account when the distributions were initially produced (see Chapter 7, § 7.5). The places, or cells, are hexagons produced

[60] See Sarkar et al. (2000) for further discussion of this example.

as part of the U.S. Environmental Protection Agency's Monitoring and Analysis program (EMAP). Each hexagon is about 649 sq. km (ignoring local topographic detail, which will, however, become important when this example is used in Chapter 7, § 7.3). A target of representation of ten was set for each vertebrate species. Obviously, these cells are far too large to be incorporated into national parks. The most that such an analysis can achieve is to target cells for conservation action, leaving the nature of the action to be decided at a local scale, through expert advice and so on. Here, however, even such a limited use is not being envisioned: the analysis has been performed only to illustrate the methodology involved.

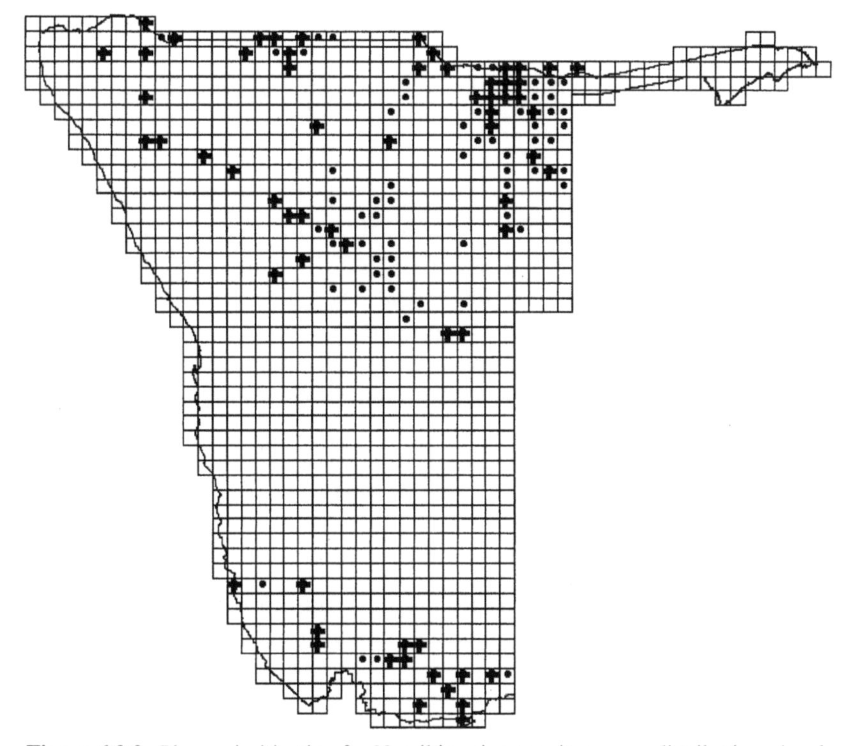

Figure 6.2.2. Place prioritization for Namibia using termite genera distributions (modified from Sarkar et al. [2002], p. 342). The data set is discussed in the text. Targets of representation were set at 100 percent for each surrogate. A goal of 5 percent and 10 percent of the area of Namibia was set. The prioritization used rarity first and complementarity second during the iterative step. Initialization was by rarity. Cells with crosses are those that are selected with a goal of 5 percent. Cells with dots are the additional cells that are selected when the goal is increased to 10 percent.

Figure 6.2.2 shows the results of place prioritization for Namibia using thirty-three genera of termites as surrogates.[61] These are presence-only data and all that are available. Data were recorded only along roads. Consequently, the data set follows the roads of Namibia. This is a very typical type of bias seen in presence-only data. Each cell is 0.25° × 0.25° of longitude and latitude. Thus the area decreased from 738.5 sq. km to 676.7 sq. km southward. As a result, these areas may be too large for selection as reserves, though, since much of Namibia is largely unpopulated and desert, some of them could potentially be so designated. However, the bias in the data precludes any serious use of these results in practical conservation planning. In this example, places were prioritized using the stated biodiversity target until a maximum area was reached for the designated places.

6.3. THE SURROGACY PROBLEM[62]

Biodiversity surrogates are features of places that enter into place prioritization procedures. Thus, in order to prioritize places, the distribution of biodiversity surrogates must be known. Surrogacy is a relation between an estimator variable and an objective or target variable.[63] Two problems must be solved: (i) a relatively theoretical one – what is to be measured?, and (ii) a practical one – can the required data realistically be collected?[64] The former is the problem of *quantification*; the latter is the problem of *estimation*. Jointly, these two problems comprise the problem of *assessing* biodiversity.[65] Finding adequate surrogates is particularly difficult – and important – when the biodiversity of a region, often on a continental scale, must be rapidly assessed, often within a year or two, because extensive changes in land use are planned. Typically, the data available will be no more than satellite images that allow vegetation cover and type to be assessed; modeled climatic variables, such as annual mean, low, and high temperatures and precipitation; presence-only records for a few species; and a digital elevation model.[66]

[61] See ibid. for further discussion of this example.

[62] For more technical details of the issues discussed in this section, see Sarkar (2004) and Sarkar et al. (2005).

[63] Note that this use of "target" differs from the use of "targets" of representation – in order to avoid potential confusion, "objective" will be preferred to "target" in the ensuing discussion.

[64] See Williams and Humphreys (1994) for this formulation.

[65] This terminology is due to Sarkar (2002) – see also Sarkar and Margules (2002).

[66] See, for example, Nix et al. (2000) for such an assessment for Papua New Guinea. The problems encountered, and the methodologies required, are particularly well laid out in that report.

Past discussions of the surrogacy problem have suffered because an important distinction – between "true" and "estimator" surrogates – has routinely been ignored. Surrogacy is a relation between an estimator variable and an objective variable. Ultimately, the intended objective of surrogacy determination is "general biodiversity," which, unfortunately, is impossible to define precisely, as will be discussed in detail in section 6.5: no matter what definition is proposed, some aspect of biodiversity will be left out. The "true surrogate" is what is taken to represent general biodiversity (its intended objective). An element of convention necessarily enters into its choice because of this problem of defining general biodiversity. Endangered and other species at some serious risk of extinction are among the most common true surrogates that have been used in practice. The justification for this choice is that it should be uncontroversial that these species form a component of biodiversity that deserves special attention.[67] Even when endangered and other species at risk are explicitly used as true surrogates, it is often implicitly assumed that all species are the true surrogates. This amounts to defining biodiversity as species diversity (see below). It should be clear that complete distributions of all species for any region will never be available: even leaving aside microbial species, complete distributions of many other taxa, including insects (which are believed to show more diversity than any other taxon), are never available in practice. Other candidates for true surrogates that have sometimes been suggested include characters or traits, life-zones, and environmental parameters.[68] However, character or trait diversity is too vague to solve the quantification problem, and environmental parameter diversity is intuitively unappealing because it is manifestly nonbiological.

In contrast to true surrogates, "estimator-surrogates" have true surrogates as their objective. Given that a true surrogate set has been precisely delineated (if only by convention), whether an estimator-surrogate set adequately represents the true surrogate set is an empirical question. Estimator-surrogates must be used because complete distributions of true surrogates are often unavailable, cannot easily be collected, and almost always cannot even be rapidly assessed. There are at least six plausible candidates for estimator-surrogates: (i) environmental parameter composition, (ii) soil type composition, (iii) dominant vegetation composition, (iv) life-zone composition, (v) subsets of species composition, and (vi) subsets of genus or other

[67] See, for example, Garson, Aggarwal, and Sarkar (2002).
[68] See Sarkar (2002) and Sarkar and Margules (2002) for further discussion.

higher taxon composition.[69] All of these have been used in practice. Usually the first three are used simultaneously.[70] They have the obvious advantage of being relatively easily assessed. Thus the estimation problem has a good solution.

The empirical relation of representation between estimator-surrogates and true surrogates (the former are supposed to represent the latter) has traditionally been interpreted in two ways:

(i) The distribution of estimator-surrogates (or some other such pattern involving them) can be used to predict the distribution (or some other pattern) of true surrogates.[71] For instance, an attempt could be made to predict the distribution of a set of species from knowledge of the available habitat. The use of ecological niche models is one method that could be used to make such predictions. However, sophisticated use of mere statistical associations and other similar algorithms has often been more successful.[72]

(ii) The places prioritized using estimator-surrogates can be used to capture what is desired of the true surrogate. The simplest way to see if this happens is to use one of the systematic place prioritization procedures discussed in section 6.2. Places can be prioritized using the estimator-surrogates with some specified target. The set of selected places can then be queried to find to what extent the required target for the true surrogates has been met.[73] The two targets need not be the same: especially if the estimator-surrogates are relatively common, a larger target can be set for them in order to satisfy a lower target for the true surrogates.[74] Once again, in order to determine whether an estimator-surrogate set is adequate, a conventional decision must be made to specify the extent to which the estimator-surrogate prioritization must capture the true

[69] These "compositions" should be viewed as a list of entries for each place and not represented as summary statistics – recall the discussion of § 6.2, which showed that place prioritization procedures require such lists.

[70] If the environmental parameters include soil types, then, ipso facto, using estimator-surrogates (i) includes using (ii).

[71] For examples within conservation contexts, see Faith and Walker (1996) and Ferrier and Watson (1997). The state of the art is presented in Scott et al. (2002).

[72] For example, the GARP modeling system (Stockwell and Peters 1999), which uses genetic algorithms, is probably the most successful method to make such predictions.

[73] This method, due to Sarkar et al. (2000) and Garson et al. (2002), is an extension of the method of species accumulation curves pioneered by Ferrier and Watson (1997).

[74] For instance, Garson and colleagues (2002) found that setting a target of representation of ten for birds was a reasonable way to ensure that species at risk met a target of representation of one in Québec.

surrogate prioritization (that is, the percentage of true surrogates that must achieve their targeted representation when all the estimator-surrogates meet their target). Determining the adequacy of the estimator-surrogates in this way requires knowledge of the distribution of true surrogates for at least some places. Thus the following three-stage process of surrogate set specification is appropriate: (a) obtain the estimator-surrogates' distributions for an entire region; (b) survey a small subset of places, appropriately randomized, for the true surrogates; and then (c) compare results using true surrogates and estimator-surrogates for the surveyed regions. If the estimator-surrogate set is adequate, then use it to represent the biodiversity of that region in all further analyses, for instance, for the prioritization of all the places in that region.

The first method is known to be difficult and so far has not yielded uniformly acceptable results for all regions to which it has been applied. However, if all that is to be accomplished using the estimator-surrogates is place prioritization, which is all that the second method requires, it is unduly ambitious.

Use of the second method has become fairly routine in surrogacy analysis in recent years, though place prioritization has sometimes been carried out only informally.[75] Are there adequate estimator-surrogate sets that can be used for rapid biodiversity estimation? This remains an open question, in part because surrogacy analysis has so far received much less attention than it deserves. The hope is that environmental parameter sets will be adequate estimator-surrogates no matter which true surrogates are used, because these sets are easy to obtain. There are often ample records of features such as rainfall and temperature patterns; moreover, these can be quite reliably modeled. Figure 6.3.1 shows some results from Texas using environmental parameter sets as estimator-surrogates and vertebrate species as true surrogates, both with a target of one. The 90 percent coverage using the entire environmental surrogate set is about as successful a use of estimator-surrogates as has ever been achieved. Even if the 90 percent coverage is deemed adequate, it would be incorrect to generalize from this result to other areas or to other true surrogates. Moreover, as will also be pointed out in Chapter 8 (§ 8.2), there may be a problem with this data set: since the vertebrate species distribution was modeled using some of the

[75] See Ferrier and Watson (1997), Howard et al. (1998), Andelman and Fagan (2000), Sarkar et al. (2000), Garson et al. (2002), and Sarkar et al. (2005).

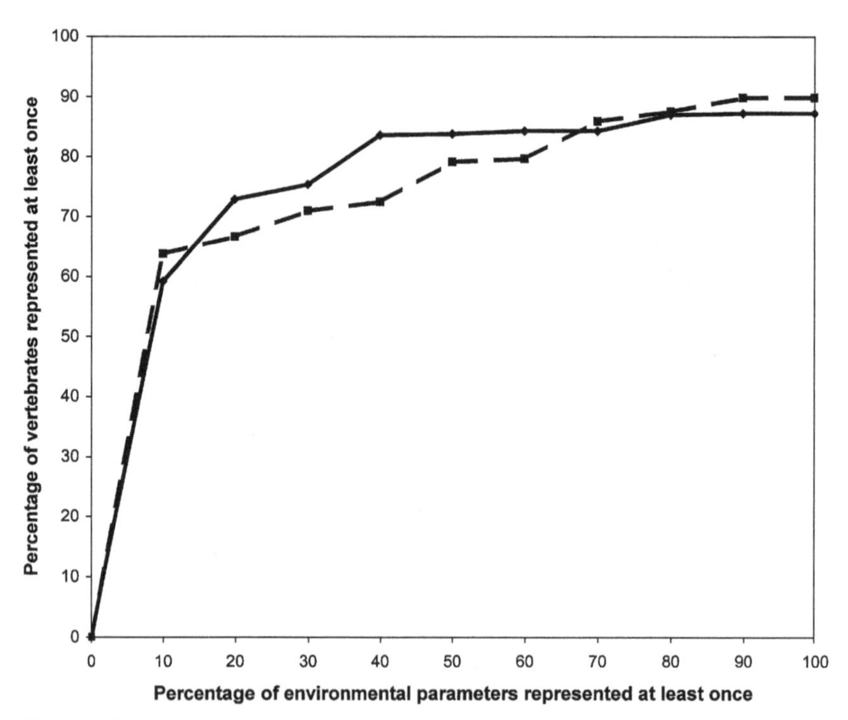

Figure 6.3.1. Surrogacy analysis using environmental data (drawn from data presented in Sarkar et al. [2000]). The percentage of vertebrate species that met a targeted representation of one as a function of the percentage of environmental parameters that met a targeted representation of one in Texas. The solid curve is the result for vegetation classes alone being used as the environmental parameter. All other subsets of environmental parameters achieved comparable results. The dashed curve is the result of using all environmental parameters available: soil types, vegetation classes, average high and low temperature, and average precipitation. The data set was described in the last section.

environmental surrogates, the good correlation may be in part a statistical artifact. If environmental parameter sets prove to be insufficient, the hope is that a more inclusive adequate estimator-surrogate set can be obtained, which supplements the environmental parameters with others that can also be easily assessed. For instance, vegetation patterns can be fairly reliably obtained by remote sensing through satellite images.

Spatial scale is also crucial to the success of surrogacy. If a place is large enough, it will contain all true surrogates and all environmental-surrogates: every feature will be a perfect surrogate for every other. If a place is small enough to hold only one biological feature, there may be no good surrogate. Is there a spatial scale at which there is no subsequent improvement in surrogacy with further enlargement? If so, then that is the appropriate spatial scale for conservation planning. The answer to this question is not known.[76] Systematic surrogacy analysis remains largely a task for the future. The most important results to date are some negative ones: traditional surrogates such as keystone species, focal species, umbrella species (those requiring large areas for their conservation), and flagship species (for instance, charismatic species around which successful public conservation campaigns can be organized)[77] are very poor surrogates for most other species. Selecting places using them is often little better than selecting places at random.[78] Once again, what is of use in conservation biology departs from the traditional wisdom of ecology, though it is hardly the case that there was much evidence that the concepts of keystone, umbrella, focal, and flagship species played any significant theoretical role within ecology. The question that demands attention, at least from philosophers of science, is why conservation biology (particularly in North America) has so often been guided by attractive intuitions rather than by systematic analyses.

6.4. THE VIABILITY PROBLEM[79]

Although a place may have a very high biodiversity content at present, it may be inappropriate to invest limited resources in conservation action directed at it. The biodiversity content of the place may be doomed for anthropogenic or other reasons. Determining the prognosis for biodiversity at a place is the viability problem: after it is solved, we have a measure of the biodiversity *value* of a place. Note that what must be established is the viability of the biological features of interest, specifically, those that are targeted for conservation, and not just the estimator-surrogates. The initial attempts to solve the viability problem in the North American context involved attempts to use island biogeography theory to determine the number of species that a potential reserve could sustain. The result was the SLOSS debate – or debacle (recall the discussion of community ecology

[76] See, however, an intriguing result of Garson, Aggarwal, and Sarkar (2002), who suggest that there is such a scale of about 330 sq. km when birds are used as estimator-surrogates for species at risk (the true surrogates) in Québec. However, recent results for Québec and Queensland using environmental estimator surrogates are less optimistic (Sarkar et al. 2005).

[77] Traditionally, these species have been implicitly but informally used as surrogates, especially in North America, without their empirical adequacy ever having been established (see below).

[78] This was shown by Andelman and Fagan (2000) at three different scales for regions in the United States, using species at risk as the true surrogates.

[79] For a more detailed account, see Sarkar (2004).

in Chapter 5, § 5.1). In the United States, the legislative context of the 1970s largely determined the course of biological discussions of viability, an exemplary case of the social determination of science.[80] The decisive event was the passage of the Endangered Species Act (ESA) in 1973, the end of a long history of U.S. federal conservationist legislation, including the Endangered Species Preservation Acts (1966, 1969) and the National Environmental Policy Act (1969). Subsequent amendments to the ESA required not only the listing of threatened and endangered species but also the designation of critical habitats and the design of population recovery plans. Other legislation mandated measures to prevent even the local extinction of species. Equally important in this context was the 1976 National Forest Management Act, guidelines for the implementation of which required the U.S. Forest Service to maintain "viable populations" of native vertebrates in each national forest.

Since populations of threatened and endangered species are generally small, attempts to implement this legislation naturally led to a focus on small populations.[81] In a 1978 dissertation, "Determining Minimum Viable Population Sizes: A Case Study of the Grizzly Bear," Mark L. Shaffer attempted to formulate a systematic framework for the analysis of effects of stochasticity on small populations.[82] Shaffer introduced the concept of the minimum viable population (MVP), the definition of which again involved conventional choices: the probability of persistence and the time period during which that probability had to be maintained were both matters of choice. One common choice was to define an MVP as a population that has a 95 percent probability of surviving for the next 100 years;[83] both numbers are a matter of convention. The conservation and recovery plans for threatened and endangered species required under the ESA (as amended) led almost inexorably to the risk assessment of small populations through stochastic analysis. The term "population viability analysis" emerged in the 1980s. That PVA based on *stochastic* models emerged from this background under what Caughley called the "small populations paradigm" should come as no surprise.[84] The optimism of the late 1970s and early

[80] Sarkar (1998b) and Kingsland (2002) appear to be the only secondary sources for this history; much more work on this fascinating topic remains to be done.

[81] Gilpin and Soulé (1986) perceptively trace the origin of the concept of the minimum viability population to the 1976 National Forest Management Act.

[82] Shaffer (1978) introduced the classification of stochasticity discussed earlier, in Chapter 5 (§ 5.1).

[83] See, for example, Hanski (1999), p. 71.

[84] Gilpin and Soulé (1986); Caughley (1994).

1980s was reflected in Gilpin and Soulé's slogan: "MVP is the product, PVA the process."[85]

By the late 1980s, it was already clear that the concept of a MVP was of very limited use, at best. Even for a single species, populations in different related habitat patches may show highly variable demographic trends, resulting in highly variable MVPs for them depending critically on the local context.[86] By the early 1990s, PVAs began to focus on estimating other parameters, such as the expected time to extinction and the probability of extinction within a specified time period. PVA is somewhat unique among the procedures used in conservation biology insofar as it does rely very heavily on the concepts, theories, and techniques of traditional ecology as it tries to predict the fate of populations. If PVA proves successful in providing empirically reliable estimates for these parameters, there will be no doubt about the centrality of its role in conservation planning. Indeed, it would justify Shaffer's grandiose 1994 hope:

> Like physicists searching for a grand unified theory explaining how the four fundamental forces . . . interact to control the structure and fate of the universe, conservation biologists now seek their own grand unified theory explaining how habitat type, quality, quantity, and pattern interact to control the structures and fates of species. Population viability analysis (PVA) is the first expression of this quest.[87]

Though hundreds of PVAs have been performed, mainly because of the legislative context of the United States, a common methodology for PVA is yet to emerge. Indeed, there is ample room for skepticism about the future of PVA in conservation biology. There are at least five sources of difficulty:

(i) The potential strength of PVA lies in the fact that the same technique can be used for all populations.[88] PVA relies on only a few parameters for all populations, generally the vital rates (intrinsic growth rates, migration rates, etc.), their variability (as measured

[85] Gilpin and Soulé (1986), p. 19.

[86] Recall the criticisms made by Margules (1989) mentioned at the beginning of this chapter.

[87] Shaffer, in Meffe and Carroll (1994), pp. 305–306.

[88] For theoretical analyses, contrary to Caughley (1994), this is equally true for deterministic and stochastic models. The expected time to extinction for large populations is $T_e = -\frac{\ln K}{r}$, where K is the carrying capacity and r is the intrinsic growth rate. Experimentally, however, as Caughley correctly points out, different designs must be fitted to different large populations, whereas the same design can be used for all small populations: the variances in the vital rates are often all that has to be measured.

by their variances), and a few environmental parameters, most notably, the carrying capacity of the environment. Consequently, models for PVA are necessarily simple and ignore local contextual differences in habitat that may well be critical for the fate of a population. This objection was part of Caughley's original criticism of stochastic PVA of small populations, but it is equally compelling for PVA of large populations. The only solution seems to be the construction of context-specific, possibly individual-based models with explicit spatial structure – recall the discussion of Chapter 5 (§ 5.2).

(ii) Almost all PVAs have been restricted to single populations of single species. Suppose that a metapopulation of a species is distributed over a landscape as n local populations. How does the expected time to extinction vary as a function of n? Beyond the trivial fact that it must grow monotonically as a function of n, even the general answer to a question as straightforward as this is not known. Moreover, multi-species PVAs have almost never been performed. But in order to conserve places, and all the biodiversity at these places, without targeting only single species, metapopulations and multiple species, organized in metacommunities, must be modeled. These are probably not insurmountable difficulties, particularly using simulation, but the paucity of work along these lines shows the current limitations of PVA.

(iii) Models for PVA are generally highly sensitive to the parameters within them because of the structural uncertainty associated with them. Yet these parameters, even such common ones as the carrying capacity, are generally recalcitrant to accurate estimation in the field, as already noted in Chapter 5 (§ 5.1).

(iv) There is considerable (structural) uncertainty about the models used for PVA. Apparently slight differences in assumptions and techniques routinely lead to widely divergent predictions. This can be illustrated using a particularly well-studied example, the population of grizzlies (*Ursus arctos*) in Yellowstone National Park, originally studied by Shaffer in his pioneering dissertation on PVA.[89] In 1994, Foley constructed a model for this population incorporating environmental stochasticity alone and depending on the intrinsic growth rate of the population and the carrying capacity of the environment.[90] The model's prediction augured well for the grizzlies: with a reasonable value for the carrying

[89] Beissinger (2002), pp. 6–7, notes the historical importance of this population in initiating PVA.
[90] Foley (1994).

capacity and the measured value for the intrinsic growth rate, the expected time to extinction was about 12,000 years. In 1997, Foley constructed another model incorporating both demographic and environmental stochasticity, but with the option of setting either of them equal to zero.[91] When this model is solved with the demographic stochasticity equal to zero, it should give the same result as the 1994 model. It does not. It predicts a much shorter time to extinction, about fifty years.[92]

(v) Finally, there are nonbiological risks that obviously affect the fate of the biological entities that are of concern. These include risks due to human activities, sometimes, as with climate change, on a global scale. Ultimately, the aim of PVA is to produce estimates of risk that take all of these other factors into account.[93] Models that attempt to do so are necessarily complex, bridled with uncertainty, and difficult to assess.

Coping with uncertainty will be one of the two foci of Chapter 7. Perhaps all that can be suggested at this stage is that, minimally, in order for the results of PVA to be used in the field, some kind of "robustness" analysis is necessary using as wide a variety of models as possible.[94] If, given two populations, all PVAs report that one has a longer expected time to extinction than the other, that is a robust conclusion, and the former should be prioritized over the latter. Similarly, if, given two management options, all PVAs report that one produces a better prognosis for a given population than the other, then that is the one that should be adopted. Beyond this, at present, it is hard to have much confidence in PVAs. It remains a field of active research, and perhaps more reliable results will emerge in the future.

PVA is not the only method that can be used to assess the viability of a candidate place for conservation action. There are at least two other options, both of which merit further development and philosophical scrutiny. These will be mentioned only briefly here. (i) Habitat-based viability analyses attempt to predict the future of populations on the basis of habitat

[91] Foley (1997).

[92] Sarkar, unpublished results. For more discussion of structural uncertainty, see Chapter 7, § 7.4.

[93] See Lacy and Miller (2002).

[94] The basic idea of this kind of robustness analysis is due to Levins (1966) and Wimsatt (1987). It has been criticized by Orzack and Sober (1993); for a response, see Levins (1993). Its formal epistemological status, whether there is a statistical sense in which this kind of analysis adds weight to a conclusion, remains uncertain, though, intuitively, the idea is obviously very appealing.

characteristics such as size and availability of resources.[95] The trouble with these is that animal species, for which such analyses are most commonly attempted, are usually structured in "clumped" metapopulations; consequently, the numbers predicted from habitat availability are almost certainly overestimates. (ii) Risk-based viability analyses assess the threat posed to a place by both anthropogenic and other factors. Such analyses are undoubtedly important, and in many situations critical, because the nonbiological – that is, anthropogenic – threat to some biological entity may far outweigh any ecological problem it is encountering. As yet there is no single framework (beyond the use of simple "rules of thumb": watch out for developers, loggers, Republicans, and so on) for carrying out risk-based analyses. But they will probably form a significant part of the conservation biology of the future, underscoring the fact that adequate biodiversity conservation will require much more than traditional biology. Insight from the social sciences will be essential.

Once the viability of biodiversity at a place has been ascertained, the list of places prioritized by biodiversity content can be refined to reflect the biodiversity *value* of a place. This procedure is implicitly carried out when, for instance, places that are already developed or have very high human densities are removed from further consideration. More systematic attempts have yet to be carried out because of the difficulty, mentioned earlier, of determining the prognosis for all biological features of interest. Once places with low viability are excluded, the place prioritization procedure can be repeated; places are now ordered by biodiversity value and can be targeted for conservation action, such as protection from changes of habitat.

6.5. DEFINING "BIODIVERSITY"

"Biodiversity" has not yet been defined in this book, even though it has been systematically used. Moreover, the claim made earlier (in § 6.3) that true surrogates for biodiversity can be chosen only by convention implicitly assumes that no satisfactory definition of "biodiversity" is forthcoming – this issue will merit further discussion here. What makes the definition of biodiversity difficult is that the biological realm – entities and processes – is marked by variability at every level of complexity. Restricting attention

[95] The publication of Manly, McDonald, and Thomas (1993) sparked systematic attempts to assess viability in this way. For an example of habitat-based viability analysis, see Boyce, Meyer, and Irwin (1994).

to entities,[96] two different hierarchical schemes are standardly used for biological classification:[97] (i) a spatial (or generalized ecological) hierarchy starting from biological molecules and macromolecules, through cell organelles, cells, individuals, populations and meta-populations, communities, ecosystems (communities and their physical habitats), and extending ultimately to the biosphere; and (ii) a taxonomic hierarchy from alleles through loci, linkage groups, genotypes, subspecies, species, genera, families, orders, classes, phyla, and kingdoms.[98] (Many intermediate levels are ignored in this description.)

There are two points to note about either hierarchy: (i) it is not clean, in the sense that biological entities fall into place in an exceptionless, operationally well-defined fashion; (ii) there is heterogeneity, responsible for biological diversity, at every level of organization. The second point is almost trivial, and a few examples will suffice to delineate its scope: almost any two populations, even of the same subspecies, will differ in their allelic profiles; except for some clonal organisms, almost any two individuals of the same species will have different genotypes; there are virtually no identical ecological communities anywhere, and so on. The first point is equally important: while some entities, such as organelles and cells, are reasonably well defined, examples such as fungi, symbionts, and clonal organisms show that even "individual" is not always precisely defined.[99]

Asexual species are notoriously hard to define, and even sexual species, usually defined by the ability to interbreed with fertile offspring, present

[96] Concern for processes leads to arguments for the conservation of biological "integrity" rather than diversity and is beyond the scope of this book; for a critical discussion, see Karr (1991) and Angermeier and Karr (1994). This issue will be considered again in Chapter 8 (§ 8.2), where it will be suggested that it merits further analysis.

[97] The philosophical importance of distinguishing between these two hierarchies is argued for in Sarkar (1998a), where it is shown that it can shed significant light on the units-of-selection controversy in evolutionary biology. It shows why alleles and individuals can simultaneously be units of selection, contrary to the popular philosophical view that selection on one of these two units is in conflict with selection on the other. Here it is similarly useful because it shows that focusing simultaneously on units of conservation in the two different hierarchies – for instance, on alleles and ecosystems – need not be inconsistent, whereas focusing *entirely* on different units in the same ecosystem – for instance, species over alleles – is much more problematic.

[98] Presumably, both hierarchies reflect evolutionary history and are constrained by evolutionary mechanisms. Since conservationist practice should take full cognizance of operative evolutionary processes, understanding the relationships between phylogeny and these two hierarchies is, in principle, critical to the design of conservation regimes. In practice, we almost never know these relationships fully but must proceed anyway.

[99] Individuality, as Buss (1987) has persuasively argued, is itself a gradually evolved phenomenon that, therefore, is a matter of degree. Consequently, the porosity of the category "individual" is hardly surprising.

problems.[100] The most striking problem is the existence of "ring species." In the United Kingdom, the herring gull (*Larus argentatus*) is easily distinguished – on morphological as well as reproductive grounds – as a species separate from the lesser black-backed gull (*Larus fuscus*).[101] However, as we go east, beginning with the Scandinavian countries and continuing around the North Pole, we find different subspecies of the herring gull, each of which can interbreed with the one (geographically) preceding it. Ten such subspecies are found if we traverse Siberia, cross the Behring Straits, and continue through Alaska and Canada. The terminal subspecies in Britain is *Larus fuscus*, which does not breed with *Larus argentatus*. Thus the usual (biological) definition of species turns out not to be transitive.

Conserving biodiversity, and construing the term intuitively to refer to all the biological diversity that there is, at every level of both hierarchies, amounts to saying that "biodiversity" refers to all biological entities. "Biodiversity" in effect becomes all of biology. Conservation would be an impractical proposal if "biodiversity" were construed in this way: everything biological would become a goal of conservation. The standard move at this stage is to suggest that three entities capture what is important about biodiversity: genes (alleles), species, and ecosystems. As a simplifying proposal in the face of intractable complexity, this convention has merit. If we conserve allelic heterogeneity completely, we take care of much of the diversity below the genotypic and individual levels of our two hierarchies.[102] If we conserve all species, we do conserve all entities at higher levels of the taxonomic hierarchy, though we may not conserve interspecific hybrids, which, because of the leakage in our classificatory schema, may not qualify for conservation. If we conserve all ecosystems, we may conserve many communities, and so on.

Nevertheless, even this catholic proposal falls afoul of the diversity of biological phenomena and does so in a rather spectacular manner. The monarch butterfly, *Danaus plexippus*, has two migratory populations in North America. Beginning in late August, the eastern population migrates to Mexico for five months. These butterflies aggregate in millions in the high-altitude fir forests in the Sierra Transvolcanica, some 80 km west of Mexico City. There are nine other such overwintering sites, all within an area of 800 km.^{2} on isolated mountain ranges between 2,900 and 3,400 m.

[100] See Sokal and Crovello (1970), reprinted in Ereshefsky (1992).

[101] Details are from Maynard Smith (1975), pp. 212–213.

[102] Much, but not all, unless we endorse a global genetic reductionism, that is, unless we espouse the view that all biological features are, in some significant way, reducible to the genes. For arguments against this position, see Sarkar (1998a).

Throughout the winter, they remain sexually inactive. Survivors migrate north, starting in late March, and lay eggs on milkweed (Asclepias sp.) along the Gulf coast. These eight-month-old remigrants die, but their offspring continue migrating north toward Canada. Two or three more generations are produced over the summer. By the end of summer, the last summer generation enters reproductive diapause and instinctively begins a southerly migration toward Mexico. The western population shows similar behavior, migrating to about forty known overwintering sites in California. What is striking about this behavior in both populations is that the migratory instinct is hereditary and yet so specific.

In California some measures have been taken to protect overwintering sites, but the sheer cost of real estate may result in only a very few of the sites getting the necessary protection. The future of the Mexican sites may be even more bleak. Though until recently the high-altitude fir forests of Mexico had for the most part been spared adverse anthropogenic effects, they now face at least six threats:[103] (i) large-scale legal and illegal logging for timber and firewood; (ii) village expansion up the mountains; (iii) increased use of fire to clear land; (iv) invasion of the forests by lepidopteran pests; (v) spraying of *Bacillus thuringiensis*, an organic pesticide, the effect of which on monarch butterflies is unknown; and (vi) increased tourism. The disappearance of overwintering sites will not necessarily mean the extinction of monarch butterflies: there are numerous nonmigratory tropical populations. However, what will disappear is the remarkable migratory behavior of the two populations just discussed, which has come to be seen as an example of "endangered biological phenomena (EBP)."[104]

Other examples include the seasonal migrations of wildebeest (*Connochaetes taurinus*) in southern Africa and the synchronous flowering of bamboo in India. While the former certainly are spectacular, and in danger of disappearing because of the construction of fences along migratory paths, the latter is perhaps even more peculiar.[105] One bamboo species, *Thrysostachys oliveri*, flowered in Burma in 1891 and seeds were sent to Kolkata and Dehra Dun, about 1,500 km apart. Clumps raised at both these places flowered simultaneously in 1940. In 1961, there was simultaneous flowering of Muli bamboo (*Melocanna baccifera*) in Assam and Dehra Dun, also about 1,500 km apart. In 1970–71, there was simultaneous flowering of spiny bamboo (*Bambusa arundinacea*) throughout India after a lapse

[103] See Brower and Malcolm (1991).
[104] See Brower's contribution to Meffe and Carroll (1994), pp. 104–106.
[105] Details are from Bahadur (1986).

of forty-five years. Clearly, a very precise biological clock exists in these species. In these cases, extinction of the species would also lead to the extinction of the phenomenon of synchronous flowering. More interesting in this context is that in extended habitats consisting of forests of a single bamboo species, flowering occurs in waves, starting at one end and ending at the other. This is the phenomenon that would disappear if these habitats were to disappear, even if the species persisted elsewhere (for instance, as isolated stands in botanical gardens). Protecting the holy trinity of genes, species, and ecosystems will typically not save such phenomena.

Even leaving aside the question of the theoretical definition of biodiversity, these recalcitrant examples give us reason to wonder whether even an adequate operational definition of biodiversity can be devised. Such a definition would merely have to justify the practices required under the consensus framework for biodiversity conservation through adaptive management elaborated earlier (§ 6.1) The proposals put forth here will not provide a fully adequate solution. Instead of adopting any of the possibilities emerging from the hierarchies mentioned earlier, the position taken here will be that biodiversity should be (implicitly)[106] operationally defined as what is being optimized by the place prioritization procedures that prioritize all places on the basis of their biodiversity content using true surrogates.[107] Thus biodiversity is the relation used to prioritize places. This definition is straightforwardly operational: it is implicitly defined by a procedure already in place. However, it does not give a measure of biodiversity as an absolute (that is, nonrelational) property of places. Rather, this definition relativizes biodiversity in two ways: (i) the definition says only whether place A has higher (or the same or lower) biodiversity than B, and (ii) it does so only against a background set of places Π that have already been targeted for protection. What the place prioritization procedure achieves is this: given Π and a set of places, it prioritizes these places on the basis of biodiversity by using the surrogate lists for Π and for each place. Consequently, place prioritization – and the rest of adaptive management – do not require an absolute definition of biodiversity. This is a deflationary

[106] The sense of "implicit" here is the technical (axiomatic) one: a set of axioms implicitly define the concepts occurring in them. (Implicit definitions are to be contrasted with explicit definitions, usually introduced by necessary and sufficient conditions.) In our case, however, we will have a variety of algorithms rather than axiom sets.

[107] For a fuller defense of this proposal, see Sarkar (2002) and Sarkar and Margules (2002). Wood (2000), p. 43, seems to be unaware of the possibility of implicit definitions when he argues that any such attempted definition of biodiversity – in his case, he has the indices of diversity in mind (see § 5.1 for an example) – must be circular.

account of biodiversity. It will probably be unsatisfactory for those who want more, perhaps an essentialist account of what biodiversity *is* in terms of necessary and sufficient conditions. Because of the heterogeneity of the two hierarchies introduced earlier, it is unlikely that any such plausible account is forthcoming. In the meantime, this account has the advantage of providing an operational definition of biodiversity for use in the practice of conservation biology.

The definition given in the last paragraph was neutral with respect to the exact and heuristic algorithms discussed in section 6.2. The rest of this discussion will assume that the heuristic algorithms are the ones that are most relevant because of their transparency. Recall that the discussion of heuristic place prioritization procedures in section 6.2 invoked not a single algorithm but a variety of related algorithms. It follows from this that different concepts of biodiversity are implicitly defined by these algorithms. The crucial rules used were rarity and complementarity: focusing on rare and new surrogates naturally captures the intuition of bio*diversity*. That rarity and complementarity should have some role in our concept of biodiversity should not be controversial. Recall that the context is one of biological conservation. We want to protect what is threatened by extinction: by and large, rarity captures that intuition.[108] We want to protect as many different entities that are not already adequately protected as possible: complementarity exactly captures that intuition.[109] It is far less clear that we are still defining "biodiversity" when we invoke the other rules – such as adjacency, area, cost, and richness – in our algorithms. It is left an open question whether "biodiversity" should be restricted to what is defined by algorithms that invoke only rarity and complementarity, or whether we should be more catholic in our taste.

Finally, endangered biological phenomena present problems not only for the definition of biodiversity but also for the place prioritization procedure itself. The only way to introduce these into the procedure is to treat the occurrence of endangered biological phenomena as a true surrogate itself. Such a move is not entirely ad hoc: since such phenomena are of obvious biological interest, it makes sense to include them in the range of biological features we would want to consider and protect, so as to understand them better. After all, from the perspective of this book it is such interest

[108] However, it does not capture it completely: some species, for instance, may naturally be rare without being in danger of extinction.

[109] Thus, this definition captures the intuition that biodiversity consists of the "differences among biological entities" (Wood 2000, p. 40).

that provides biodiversity with the transformative value that obligates us to attempt to conserve it. Nevertheless, it is also not entirely satisfactory: so far, the other biological features that have been considered have all been, for want of a better term, structures. In contrast, endangered biological phenomena are processes. Consideration of processes does not enter very naturally into discussions of biodiversity conservation. There will be more discussion of this point in Chapter 8 (§ 8.2).

7

Incommensurability and Uncertainty

Suppose we have at hand a set of lists of places that have been prioritized by biodiversity value. Presumably, each member of this set was initially generated by prioritizing for biodiversity content using the methods described in the last chapter. This list was then filtered to take into account the prognosis for biodiversity at each place, resulting in a list prioritized by biodiversity value. It now remains to incorporate the other (for instance, anthropocentric) considerations that must be taken into account during biodiversity conservation planning.[1] If biodiversity conservationists were the only contestants in the struggle over land, nothing more than prioritization by biodiversity value would be necessary. Places would be targeted for conservation on the basis of the prioritized list, perhaps constrained globally by cost and similar considerations.[2] Unfortunately, there is no dearth of other contestants for land.[3] These contestants evaluate our lists of places according to criteria other than biodiversity.[4] In the following discussion (up to the end of

[1] The methods discussed in this section can also be used to incorporate design criteria (see Chapter 6, § 6.1) that affect viability. The discussion will be focused on anthropocentric criteria, such as economic and social cost, only for expository simplicity.

[2] Once again, for expository simplicity it is being assumed that a biodiversity conservationist is only interested in biodiversity *qua* biodiversity conservationist. An individual may, of course, both be a biodiversity conservationist and share the other concerns that are being referred to here.

[3] The conflict being presented in the text as an interpersonal one (once again, for expository simplicity) can also occur as part of the decision-making process of a single individual attempting to incorporate multiple criteria into decision making. Indeed, that is how it has usually been studied. The methods discussed in this chapter are applicable to both cases.

[4] Even if it were possible to attribute intrinsic$_2$ value to the relevant features of biodiversity (for instance, species or species assemblages), conflicts of value between criteria would still have to be negotiated, a point that is often missed by advocates of attributing intrinsic value to all features of biodiversity. An example will help to clarify this point: suppose, not unreasonably, that intrinsic$_2$ value is attributed to both human freedom and human life. These values will be in conflict if there is credible evidence that one individual intends to murder another. In such a circumstance,

section 7.3), these other criteria will sometimes be said to reflect other "values"; this use of "value," though common in environmental philosophy, is strictly distinct from the use of "value" in the preceding chapters (except Chapter 4, § 4.1), though the two uses are consistent with each other: instead of saying that there are different criteria that each attributes value to alternatives, it is a useful shorthand to say that there are different values associated with an alternative.[5] (Thus, for instance, according to this use of "value," we can call biodiversity one of our values.)

These other criteria not only include economic well-being – in particular, the ability to provide the basic necessities of life (such as food and shelter) – but also reflect "values" such as an attachment to traditional habitats or practices, or even to wilderness. These criteria are often *incompatible* in the sense that the satisfaction of one leads to the decline of another. Throughout this chapter, when biodiversity is being considered along with other criteria, it will be assumed that not all of these criteria are compatible with each other. This is not a controversial assumption: it is clear that economic utilization is often incompatible with biodiversity conservation. It is also incompatible with wilderness preservation. As noted in Chapter 2 (§ 2.3), biodiversity conservation is sometimes incompatible even with wilderness preservation.

Sections 7.1–7.3 of this chapter are about navigating these diverse values (as reflected in divergent evaluative criteria). In particular, consistent with the approach taken in the last chapter, there will be an emphasis on systematic (and, as far as possible, algorithmic) approaches to mediate such conflicts of value. In the literature of policy analysis and decision making, where the problem is most often studied, this problem is usually called the "multiple criteria optimization problem" (or "multiple criteria decision-making problem"). That terminology will not be adopted here, because the only solution that will be defended, on some very general philosophical grounds, will not constitute "optimization" in any ordinary mathematical sense of the word. Instead, the problem will be called the "multiple criterion synchronization problem," the underlying idea of the proposed solution being to synchronize incompatible values to the extent possible.

Sections 7.1–7.3 should not be construed to suggest that all such conflicts of value – or even the more important of them – can be negotiated using

there would be full justification in curtailing the freedom of the former, for instance, through preventive detention.

[5] As Keeney and Raiffa (1993, p. 67) put it, in the context of a discussion of multiattribute value theory: "We will unashamedly use the same symbol X for the attribute in question and the evaluator of that attribute." These attributes correspond to our criteria; the evaluator returns a value of that attribute for each alternative.

the algorithmic techniques discussed here. As the discussion of section 7.1 will emphasize, some values are even intangible (or, equivalently, the corresponding criteria do not allow the alternatives to be successfully ranked in a linear order). These may well be the most important of our values, and they may be beyond the reach of our algorithmic methods during conservation planning.

Sections 7.4–7.6 of this chapter will be concerned with the problem of making rapid decisions under uncertainty. There are at least four relevant sources of uncertainty during decision making in the context of biodiversity conservation:[6] (i) the structural uncertainty of the best available models, (ii) environmental uncertainty, (iii) partial observability, and (iv) partial controllability. (Recall, also, the discussion of Chapter 5, § 5.2.) It will also be argued (in § 7.6) that the framework of Bayesian statistics provides some hope for coping with the least tractable sources of such uncertainty, even though the solutions that can so far be offered leave much to be desired.

Throughout this chapter, the discussions will be framed with the problem of prioritizing places for conservation action as the common background. The discussions of multiple criteria and of uncertainty both raise a host of technical and philosophical problems, all of which merit much further exploration than what has so far been attempted in either the philosophical or the scientific literature. Both parts of this chapter concern topics on the frontiers of research in the environmental sciences: they are not definitive treatments of their subjects and should be regarded as an open invitation to criticism and further research by readers.

7.1. TYPES OF VALUES

Recalling the discussion of Chapter 4 (§ 4.1), two values are *commensurable* if they can be assessed on the same ordinal scale, that is, if alternatives with these values (for instance, places to be prioritized) can be ordered on the same scale;[7] otherwise, they are *incommensurable*. Ordinality is a weaker condition than quantifiability. Let α, β, and γ be three different points on this scale. All that ordinality requires is that α, β, and γ can be placed in a linear order, for instance, (α, β, γ), with no indication of the extent of the difference between α and β when compared to that between β and γ. Quantifiability requires, further, that a definite numerical value be assigned

[6] Williams (1997, 2001).

[7] Equivalently, alternatives can be ordered using the corresponding criteria on the same scale.

to each of the entities α, β, and γ. (In both cases, note that two alternatives may have the same value, that is, be represented by the same point on the scale.) For instance, the market allows apples and oranges to be ordered on the same scale; in fact, in this case market mechanisms also allow them to be put in the same quantitative scale of monetary value. Suppose that α is harder than β if and only if α can be used to scratch β. Hardness can be used to put all solid materials on an ordinal scale. However, this order cannot (or at least cannot easily) be quantified.

Conventional economics assumes not only that all values are commensurable, but that they can be put on the same quantitative scale of utility, for instance, monetary value in the case of apples and oranges just discussed.[8] It is easy enough to see how utility as measured by monetary value may be estimated for the amount of timber that may potentially be logged from a place. This is the cost of "forgone" opportunity if the place is put under conservation rather than logged. Less convincingly, an attempt may be made to attribute a monetary value to the tourism at a place: for instance, this may be measured by the estimated expected profit of opening up that place to tourism. Estimates of this sort are notoriously problematic because of the empirical problems associated with them. Human behavior such as tourism is far too dependent on a wide variety of highly contextual and contingent factors for the precise values of such estimates ever to have much reliability. Nevertheless, there is, in principle, no telling objection to putting the logging value and the tourism value of a place on the same monetary scale.

Conventional economics has a medley of techniques – of varying degrees of plausibility – to make these assessments of the value of features such as biodiversity. A critical assessment of all of them is beyond the scope of this book. To motivate a transition to less conventional strategies, it will be sufficient here to note the most general problems in the context of the strategy that is most commonly proposed, though hardly ever practiced: assessing the willingness to pay (WTP) of individuals in order to assess the utility of some entity.[9] (There are more complex methods, but none entirely escapes the problems mentioned here in the sense that they guarantee the provision of a fully satisfactory measure.) Note, first, that any such strategy can potentially receive two types of justification: (i) empirical, where it is shown from data that the strategy leads to consistent and plausible attributions of utility to

[8] Recall the discussion in Chapter 4 (§ 4.1). Note, however, that while utility is a monotonic function of monetary value, the two are not the same.

[9] For a more extended discussion, see Norton (1987). Perhaps the most important technique not being analyzed is that of the valuation of ecosystem services – but see Chapter 4, § 4.1. For a critical analysis of contingent valuation techniques, see Bateman and Willis (1999).

initially unlikely candidates; and (ii) theoretical, where assumptions based on shared norms are shown to justify the strategy.

In its most basic form, a WTP assessment is supposed to begin by asking potential consumers what each of them is willing to pay to conserve the biodiversity of a place. These data – usually after significant further treatment – are then used to provide a monetary biodiversity value for that place, thereby rendering it commensurable with entities that have immediate demand values. At the empirical level, the first – and critically important – point to note is that this procedure has very rarely, and with only partial success, been carried out where the assessment of monetary biodiversity value (or that of wilderness or any other such criterion) is of practical consequence.[10]

Moreover, there are at least three (related) reasons to doubt that, empirically, this procedure is likely to produce acceptable results. The discussion here gives a fuller treatment of one of the objections to the use of demand values that was mentioned in Chapter 4 (§ 4.1):

(i) Responses to WTP questions are often likely to be meaningless. Most individuals have no idea what they are willing to pay for the conservation, say, of the Big Bend National Park in Texas, though they may clearly not be willing to see it transformed into a set of ranches. This is simply not how they think of Big Bend, in terms of such a monetary value. They may even be willing to be politically active to save Big Bend (were it under any serious threat). They may be willing to contribute something toward its upkeep, perhaps to pay more for specific conservation projects. But these monetary figures say little about what they are willing to do, including what they are willing to pay, to save Big Bend from impending destruction. Are they willing to pay more for Big Bend than, say, the Big Thicket National Preserve (also in Texas)? Most individuals simply do not know. Any answer given to a WTP questioner will be little more than arbitrary; different responses will almost always be generated depending on how the relevant decision procedure is presented to respondents.

(ii) If there is more than one WTP question, the answers may not aggregate consistently. Suppose the WTP questioner wants to find the WTP data

[10] Even without the restriction to practical consequences, such valuations seem never to have been systematically carried out to assess the plausibility of the procedure in the field. The most comprehensive effort in this context is that of Balmford and colleagues (2002), which was discussed in Chapter 4 (§ 4.1). Even it is only partially successful, because it attempted to show only that nondestruction of the selected habitats was economically preferable to their destruction.

for a set, Π, of n separate places, π_i ($i = 1, \ldots, n$) and suppose the questions about each are posed sequentially, perhaps over a significant period of time.[11] Suppose an individual's response to the WTP question for π_i is X_i. (For the sake of argument, suppose that this answer is meaningful in spite of [i] above.) This means that the individual is supposed to be willing to pay $X = \sum_{i=1}^{n} x_i$ for the entire set Π. In most circumstances this estimate is likely to be wildly implausible. Suppose that, when asked, an individual is willing to contribute 5 percent of her or his annual income to protect Big Bend. Later, the same individual responds with 7 percent in the case of Big Thicket. But this should not be interpreted as saying that the individual in question is willing to contribute 12 percent of that individual's income to save both Big Bend and the Big Thicket. For most individuals, 12 percent of annual income is far too much to contribute to protect national parks and reserves, no matter how important for biodiversity or wilderness they may be. At a more abstract level, the choices posed to individuals by WTP questions are not independent of each other: what the individual may be willing to pay for one option depends on what other options will also require payment.

For conventional economists, there is one potential way out: WTP questions should be posed only for complete sets of (potentially mutually interdependent) options, and thus the nonindependence of these options will automatically be taken into account. The trouble is that this strategy is almost impossible to carry out in practice. In the context of biodiversity conservation, not only all places that are currently of interest but all places that are going to become of interest in the projected period during which an individual may be expected to make payments must be known. It should be clear that this requirement can never be met in practice; more importantly, it does not seem plausible that it could even be approximated.

(iii) Responses to WTP questions may not reflect the respondent's utilities. An individual may regard Big Thicket as being more valuable than Big Bend (for instance, for its higher biodiversity content) and yet be willing to pay more for the conservation of the latter than for the former

[11] In practice, it is quite likely that a significant period of time will be involved. Different places come under threat at different times. A potential developer may begin to eye Big Thicket only when it becomes clear that Big Bend is not for sale. Consequently, the WTP question for Big Thicket would be posed long after the one for Big Bend, possibly so much later that an individual would have forgotten the answer given to the first question.

if there is a perception that Big Bend is under serious threat, whereas Big Thicket is not.[12] Or an individual may have a higher WTP for Big Thicket than for Big Bend, for Big Bend than for the Padre Island National Seashore, but for Padre island than for Big Thicket (provided that the relevant questions are asked only for each pair at a time). There is nothing to guarantee that the results of WTP questions are transitive. This is yet another example of the well-known problem of the intransitivity of preferences. It calls into question the very consistency of the results of any procedure based on WTP. The best solution available to an advocate of WTP strategies is to suggest that the individual preferences be revised iteratively in an effort to achieve consistency. This may work in some situations, but there is no guarantee that it will always do so.

At the theoretical level, attempts to justify such procedures fare no better. The basic problem is the justification of the assumptions (i) that all values (or criteria) can be reduced to one scale, and (ii) that this scale of utility is the same one that fixes monetary prices in the market, and that points on this scale can be estimated using the monetary values.[13] These assumptions usually form part of what is called the economic model of rationality, which is not only suspect but sometimes called into question even by conventional economists, though what is usually criticized is not these assumptions but the unlimited computational and analytic power that is attributed to agents by this model.[14] Note that (ii) is relevant only if (i) is satisfied.

[12] One may, of course, redefine utility in order to take threat into account. This is exactly what happens if threat considerations are used in determinations of viability (recall the discussion in Chapter 6, § 6.4) and viability is then used to determine biodiversity value. Consequently, this objection cannot be regarded as being on a par with (i) and (ii).

[13] That the monetary scale is the appropriate point of departure is itself far from obvious. As early as 1861, John Stuart Mill observed in *Utilitarianism*:

> There is nothing originally more desirable about money than about any heap of glittering pebbles. Its worth is solely that of the things which it will buy; the desire for other things than itself, which it is a means of gratifying. Yet the love of money is not only one of the strongest moving forces of human life, but money is, in many cases, desired in and for itself; the desire to possess it is often stronger than the desire to use it, and goes on increasing when all the desires which point to ends beyond it, to be compassed by it, are falling off. It may, then, be said truly, that money is desired not for the sake of an end, but as part of that end. From being a means to happiness, it has come to be itself a principal ingredient of the individual's conception of happiness. (Mill 1962, p. 290)

Perhaps all that has changed is that, according to conventional economics, money has become the sole measure of the happiness to which Mill alludes.

[14] For a discussion of bounded rationality, see Jones (1999).

Why should these assumptions be accepted? The standard argument is to claim (i) that WTP (and similar) assessments are necessary for conventional economic analyses and (ii) that the empirical success of the latter provides justification for the former. This argument is methodologically respectable: this is the standard strategy for the justification of most theoretical assumptions of almost any science. However, claim (ii) is obviously politically loaded: the traditional Left would deny almost any success of conventional economics, while the Right would bow at its altar without shame. Nevertheless, there should be little doubt that in many local contexts, at small temporal and spatial scales, conventional economics has often been (a) descriptively, that is to say, predictively, successful and (b) normatively useful, that is to say, to the extent that it prescribes action, the actions seems at least plausible. If it has been predicted successfully that customers at one ocean resort prefer sea-facing rooms to those that face the swimming pool, it makes sense to try to maximize the former at the expense of the latter at the next hotel to be built.

Nevertheless, this is far from the scale of success required for the justification of strategies such as WTP assessments of the demand value of biodiversity (even ignoring the fact that that the arguments of Chapter 4 show that the ultimate justification for biodiversity conservation relies on its transformative value). There is little evidence that such analyses can illuminate most important facets of human cultural and intellectual life (including the pursuit of biodiversity conservation): attempts in that direction have usually produced more hilarity (which is not without value) than insight. For instance, the average U.S. household is supposed to be willing to pay $257 to prevent the extinction of the bald eagle (*Haliaeetus leucocephalis*); given about 100 million households, the United States as a whole is then supposed to be willing to pay more than $25 billion to prevent this extinction.[15]

Moreover, it is a fairly widely held belief among biodiversity conservationists that too much respect for conventional economics – often in the form of the relentless pursuit of immediate profit – has led to the widespread destruction of natural habitats during the last few decades. From

[15] The average household is also supposed to be willing to pay $208 for the prevention of extinction of the humpback whale (*Megaptera novaeangliae*) and another $80 for the grey wolf (*Canis lupus*) (Loomis and White 1996). Thus the United States as a whole is willing to pay $125 billion for just these three species. Even though these are lump sum and not annual figures, it is hard to see how, if the procedures of conventional economics are reliable, we do not have enough money for every dream a biodiversity conservationist may have. (The raw figures were converted to 2000 dollars using the consumer price index in Heinzerling and Ackerman 2002.)

this perspective, attempts to ascribe monetary value to biodiversity constitute an effort to bring biodiversity into the orbit of conventional economics precisely because of a failure of the quest for profit to protect biodiversity. It follows that attempting to show that implausible strategies intended to treat biodiversity in this way are justified because of the *success* of conventional economics is circular reasoning.

The discussions of this chapter will not assume that all criteria or values are commensurable. Rather, this chapter will concentrate on decision making when there are incommensurable values. The fact that such methods can be developed to an extent sufficient for most contexts of biodiversity conservation underscores the relative unimportance of attempts to reduce all values to a single scale. Thus defenders of classical economic techniques do not even have recourse to a "no other option" argument that says that we need to effect such a reduction because, otherwise, no analysis of our options is possible.

Returning again to the discussion of Chapter 4 (§ 4.1): in addition to the distinction between commensurable and incommensurable values, yet another distinction was made there, between tangible and intangible values. A value is *tangible* if and only if entities bearing it can be put on an ordinal scale, that is, if places or, more generally, alternatives (see the analytic framework developed in § 7.2) can be linearly ordered according to this value; otherwise, it is *intangible*. Different tangible values may be incommensurable with each other. According to the definition just given, all intangible values are incommensurable with each other and with all tangible values.

Intangible values may arise in at least two ways, the second of which is somewhat counterintuitive:

(i) Individuals may be unwilling to rank alternatives. For instance, it is often hard to see how two sacred places should be ranked with respect to each other; nor, given what was said earlier, is there *any* compelling reason for insisting on such a valuation. One may similarly be unwilling to rank two places on the basis of their wilderness value. Conventional economics has a pat answer for such situations: the choices are supposed to be indifferent and ranked on a par with each other. There is both a practical and a theoretical objection to this move. At the practical level, a bewildering variety of alternatives – perhaps all those that are available – may become indifferent. Every place may be sacred to some cultural group. Trying to treat such phenomena within the context of conventional economic analysis may lead to an impasse and provide no help in the prioritization of places. At the theoretical level, a refusal to rank alternatives does not mean that an individual is indifferent with respect to differences between them. For instance, in the case of sacred places, a refusal to rank places may reflect deeply held values about sanctity and so on. However, even if an individual is unwilling to rank some set of alternatives, that unwillingness may well disappear in localized contexts, for instance, when the individual is presented with a choice between exactly two places, one of which is sacred and the other not. One possible response to this last situation is to require that assessments of indifference always be relativized to contexts. The trouble is that, in the face of intangible criteria, it is impossible to determine and analyze all possibly relevant contexts.

(ii) According to the definition given above, if some criterion leads to an intransitive ordering of alternatives – which means that they cannot be put on a linear ordinal scale – it is intangible. This move is counterintuitive insofar as it does not reflect the customary connotations of "intangible," which invoke the idea of an impossibility of measurement rather than of ordering. Nevertheless, the problems posed by this situation for the analysis of prioritization problems are almost identical to those posed by (i); the procedure described here will not be able to accommodate criteria that do not lead to a linear ordering of places. This problem will, therefore, be treated as part of the problem of intangible criteria.[16]

From a pragmatic managerial perspective, it may seem that, in spite of the problems discussed here, the problems posed by incommensurables and intangibles should be resolved – or at least avoided – by fiat, by simply accepting the techniques of conventional economic analysis with all its attendant baggage. This would ease decision making in the field. Moreover, we deployed a similar strategy in Chapter 6 (§ 6.5) when faced with the equally recalcitrant problem of defining "biodiversity." Similarly, in the present circumstance, we should be willing to countenance some initially counterintuitive procedure in order to frame a practical program having some chance of success. For the prioritization of places using a multiplicity of values, this argument is less than compelling for at least three reasons:

(i) Most importantly, conventional economic analyses have no established record of success in contexts such as that of biodiversity conservation. There seems to be no example where multiple values have been reduced

[16] At most, the dispute here is terminological. It can be resolved by reticulating our terminology further and introducing a new term for the problem described in (ii).

to a unitary monetary scale using WTP assessments or other techniques in a way that makes the resultant values plausible to skeptics.[17] It should be noted, though, that it is far from clear that there have been truly systematic efforts to this end. Nevertheless, the arguments given earlier suggest that the chance of success is not high.

(ii) Conventional economic analysis is not the only game in town. As section 7.2 will mention, considerable progress can be made toward place prioritization without making the questionable moves that such analysis requires. For many contexts of biodiversity conservation, a good enough solution can be found.

(iii) Part of philosophical analysis – which is the main task of this book – requires a careful analysis of which parts of a scientific framework rest on relatively solid ground (direct connection to experiment, etc.) and which parts are comparatively arbitrary (and based almost entirely on conventional choices made on pragmatic grounds). At the normative level, the latter parts should always be questioned, particularly if it seems possible to reconstruct the framework without them or with less reliance on them.

Thus, ignoring the incommensurability and intangibility of some values on pragmatic grounds and adopting the methodology of conventional economics seems unwarranted, or at least premature, before exploring other possibilities.

Since the emphasis of this chapter is on systematic methods, nothing more will be said about intangible criteria beyond this paragraph. This is not to suggest that they are either philosophically or practically unimportant – they often reflect deep philosophical principles and sometimes have practical consequences of signal importance to biodiversity conservation. As an example of the former, return to the case of the Olive Ridley turtles (*Lepidochelys olivacea*) introduced at the beginning of this book (Chapter 1). It may be plausible to regard the availability of food as a tangible criterion or value. However, if the existence of this resource is a matter of sheer survival for the local (human) community, then the value in question may well be regarded as too important for alternatives to be ranked according to it (for eventual use by an algorithm). We may refuse to rank an alternative by how many communities may survive, for the same reason that we may refuse to rank an alternative by the number of people killed or enslaved. The value now comes to be treated as an intangible while

[17] For further discussion of this point, see Heinzerling and Ackerman (2002).

nevertheless having its importance acknowledged. As an example of the latter situation, sacred groves belonging to temples have been critical to the conservation of the last remnants of the rainforests of the Western Ghats of India, one of the global hot spots of plant biodiversity.[18] These groves reflect intangible religious values. But, other than noting that intangible values are often immensely important in motivating human actions, there seems to be little more that can be said about them. In section 7.2, each (tangible) incommensurable value will be treated as a distinct criterion during multiple criteria synchronization. Thus, though intangible values are admitted into the discussion in principle, the methods described have no place for them.

7.2. MULTIPLE CRITERION SYNCHRONIZATION

The framework adopted here is essentially the one originally explored by Arrow and Raynaud in 1986;[19] some of the terminology used here is of more recent vintage.[20] However, because the goal here is see how far we can go using rational[21] assumptions and without introducing any arbitrary assumptions, much of the more recent work on this problem is not analyzed here; as will be noted later, in much of this recent work special models with arbitrary assumptions are routinely used to resolve the problem to a greater extent than purely rational assumptions would allow. It turns out that purely rational assumptions suffice for most of the situations encountered in the context of biodiversity conservation. An effort is also being made here to avoid jargon and technical terminology to the extent possible.

Multiple criterion synchronization (MCS) starts with two structures: a set of *alternatives, A*, and a finite set of *criteria, K*. An alternative is a decision such as that to conserve a particular place. Each criterion reflects a value, as noted earlier (§ 7.1). In order for all these values to remain tangible, it is required that each criterion $k_i \in K$ is able to induce a weak complete ordering $\leq_i^*$ on A, that is, for any two $\alpha_m, \alpha_n \in A$, $\alpha_m \leq_i^* \alpha_n$ or $\alpha_n \leq_i^* \alpha_m$ (or both). This requirement captures the idea that any tangible value must make it possible for all alternatives to be ranked with respect

[18] For details on sacred groves, see Gadgil and Berkes (1991) and Gadgil and Guha (1992).

[19] Arrow and Raynaud (1986). For further discussion of the term "multiple criterion synchronization," see Sarkar and Garson (2004).

[20] See Bogetoft and Pruzan (1997).

[21] There will be no attempt to define rationality here; what it means should be clear from the later discussion, where it will be taken to imply that nothing more is known about the criteria than what is explicitly indicated in the statement of the problem.

to each other. Customarily, this requirement is met by positing, for each $k_i \in K$, a function $\mathbf{f}_i$: $A \rightarrow \mathbf{R}$, where $\mathbf{R}$ is the set of real numbers. Thus, if $\alpha_m \in A$, $\mathbf{f}_i(\alpha_m)$ is a real number. Each $\leq_i^*$ then maps to the customary $\leq$ (less than or equal to) relation in $\mathbf{R}$. With the analogy of cost in mind, alternatives that are "lower" by the $\leq_i^*$ are presumed to be better in this formalism. However, two related points should be noted: (i) positing the existence of such $\mathbf{f}_i$ introduces more structure than the original definitions require,[22] and (ii) the definitions merely require an ordinal structure to be imposed by the $\leq_i^*$ on A. The $\mathbf{f}_i$ do more than that: they impose a metric structure, that is, distances between the α_m, $\alpha_n \in A$ are also defined (thus allowing the parameters mentioned in [i] to make sense). If analytic rigor is required, then care must be taken not to invoke such illegitimate metric properties introduced by the f_i.[23]

Given this structure, the basic problem, often called the "basic decision problem," that an MCS procedure must solve is to define a weak complete order on A, henceforth called a canonical order, $\leq^c$, such that, for every $k_i \in K$, it captures the $\leq_i^*$ it defines on A. Once such a $\leq^c$ is found, it becomes trivial to find the $\alpha \in A$ such that, for all $\alpha_m \in A$, $\alpha \leq^c \alpha_m$. This α is the one that is wanted: it is the best alternative using all the criteria. The sense of "capture" in defining a $\leq^c$ is left embarrassingly vague in the MCS literature. Arrow and Raynaud demand that the $\leq^c$ be a "legitimate synthesis" of the $\leq_i^*$.[24] In the same spirit, Bogetoft and Prezan demand that we find a preference function that includes all the criteria, $k_i \in K$.[25] The conditions of legitimacy are not laid out explicitly. The weakest plausible notion of "capture" is to require only that a $\leq^c$ be consistent with all the $\leq_i^*$, with consistency being captured by the requirement that for all α_m, $\alpha_n \in A$, and all $k_i \in K$,

$$\alpha_m \leq_i^* \alpha_n \Rightarrow \alpha_m \leq^c \alpha_n.$$

If the different criteria are incompatible, even such a weak $\leq^c$ cannot usually be found without introducing new assumptions that are arbitrary and, therefore, of dubious epistemological status.

The interpretation of A is that each $\alpha_m \in A$ consists of an action that may be carried out, for instance, a conservation plan that may be adopted.

[22] For instance, parameters such as $\mathbf{f}_i(\alpha_n) - \mathbf{f}_k(\alpha_n)$ and $\frac{\mathbf{f}_i(\alpha_m)}{\mathbf{f}_k(\alpha_n)}$, while obviously well defined in $\mathbf{R}$, do not have to exist in order for the explicit assumptions of the definition to be satisfied.

[23] What is being called a metric structure here is sometimes also called a "cardinal" structure.

[24] Arrow and Raynaud (1986), p. 1; "legitimate" seems to imply normativity, but the norms (if any are intended) are left entirely undefined.

[25] Bogetoft and Pruzan (1997), p. 50; the sense of "include" is left unspecified.

(Thus, what an MCS procedure returns is putatively the best action given the multiple criteria to be satisfied.) In principle, A can be infinite. In practice, there are also many other problems that may be associated with A. There may be problems distinguishing between the available alternatives that belong to A. The boundaries of A may not be precise: last-minute candidates may emerge as viable alternatives during the decision process. Because of these and other potential problems, the identification of A is often regarded as an "approximate operation."[26]

In the present context (of place prioritization for biodiversity conservation), these complications can almost always be avoided. A can be a set of two types:

(i) Each element $\alpha_m \in A$ is a set of places that forms a potential target for conservation, that is, each element of A constitutes a potential conservation area network. The goal of an MCS procedure is to select one element of A as the best potential conservation area network. Procedures of this sort will be called *terminal state* procedures. It is assumed that a reasonable set of alternatives has already been selected. MCS procedures – that is, the consideration of multiple criteria – come in only to select the alternative finally judged to be the best. In practice, this is the strategy that is used when biodiversity is privileged over the other criteria (see § 7.3). A is the set of networks of places judged to be equally preferable on the basis of biodiversity considerations alone; other criteria, such as cost and area, may then be invoked to select a member of A. The action to be carried out is the designation of the selected set of places for conservation. This is the scenario with which this chapter started – reasons in favor of adopting it will be discussed in the next section.

(ii) Each element $\alpha_m \in A$ is an individual place. In this scheme, A is the set of places, Π, invoked in the preliminary place prioritization algorithm described in Chapter 6 (§ 6.2). Given a set of selected places (initially the null set), MCS procedures are used to select the next best place according to the set of criteria; the final result is an ordered list of places prioritized according to all the criteria. Procedures of this sort will be called *iterative*. Typically, such procedures are used when all criteria, including biodiversity, are treated as being on a par with each other (see § 7.3). In this case, the action to be carried out is the inclusion

[26] See Arrow and Raymond (1986), p. 8.

of the selected place in the prioritized list of places to be preferentially targeted for conservation action.

It is important to note that, for both options, (i) A is finite. The finitude of A makes the analysis of alternatives much simpler in this context than in the more general case described earlier. (ii) Trivially, the different alternatives (sets of places or places) are distinguishable from each other; and (iii) last-minute alternatives are almost never brought into consideration – the qualification "almost" is required because decision makers may choose to reintroduce previously excluded places in iterative procedures (for instance, because of intangible but nevertheless psychologically important values). In practice, the most serious problem that will have to be faced is that it will routinely be impossible to select a unique alternative without invoking additional arbitrary assumptions.

Each criterion $k_i \in K$ is a formal (that is, algorithmic) representation of a tangible value. It has been assumed that this set is finite; an extension of this discussion to infinite sets is possible but introduces formal complexities with no accompanying new insight. Ideally, the set of criteria should satisfy four conditions:[27]

(i) *Completeness*: the set of criteria should encompass all the values that are relevant to the problem. In the present context, in principle, because of the admission of intangible values into consideration, this condition cannot be met. The best that can be achieved is the inclusion of all tangible values. In practice, even achieving this is notoriously difficult. It is obvious that there are tangible economic costs associated with conserving any place. Nevertheless, these are routinely well-nigh impossible to estimate accurately; and even simply ordering places on the basis of cost is often difficult. And economic cost is probably among the simplest of the criteria that must be taken into account. Even if all tangible values are successfully included, the acceptance of intangible values means that an MCS algorithm cannot produce the final decision.

(ii) *Minimalism*: the set of criteria should be kept as small as possible. There is an obvious practical justification for this condition: it attempts to keep the MCS procedures – and decision making in general – as simple as possible. More importantly, there is also a theoretical justification for this condition, which will be discussed in the context of the independence criterion ([iii] below).

[27] The discussion here follows, *but only to a very limited extent*, that of Bogetoft and Pruzan (1997), p. 116.

(iii) *Independence*: ideally, all criteria $k_i \in K$ should be independent of each other, that is, each of them can be changed without altering the others. In practice, ensuring genuine independence can be tricky. Recall the use of cost and area as constraints during the discussion of preliminary place prioritization in Chapter 6 (§ 6.2). It is quite likely that, in most contexts, area constraints are taken into account because of cost considerations: larger areas cost more to acquire and conserve than smaller ones. In order to achieve independence, one would have to partition the economic cost into the cost due to area and other costs; only the latter should enter K. In practice, this is difficult but not impossible. Finally, minimalism and independence together ensure that redundant criteria are not included in K.[28] This is important because redundant criteria would bias considerations in favor of some values – those corresponding to more than one criterion – over others. However, strictly speaking, for the algorithm described here, complete independence of the criteria is not formally required. All that is required is *partial* independence.

(iv) *Operationality*: every criterion $k_i \in K$ should be such that it is possible, in the field, to order the alternatives according to it. This is a stronger requirement than tangibility: as noted in the discussion of (i) earlier, a criterion may confidently be known to be tangible, yet its estimation in the field may be well-nigh impossible. There is a serious problem here: criteria that are operationally sound may not be adequate as a representation of any of the values that are germane to the problem, whereas those values that are germane may not be operationally accessible.[29] There is no general formal (algorithmic) solution to this problem. In the present context, this problem is somewhat less serious than it might be elsewhere, because of the explicit recognition of intangible values. Those criteria that truly resist empirical assessment can be included among the intangibles and dropped from consideration at this stage.

Unlike the situation with the set of alternatives, the context of biodiversity conservation does not make the identification of an adequate criterion set any easier than the general case (despite the slight advantage mentioned in [iv]).

[28] Bogeloft and Pruzan (1997), p. 166, include nonredundancy as a separate condition; this appears to be unnecessary.

[29] This has led Arrow and Raymond (1986), pp. 8–9, to distinguish between attributes ("well identified aspects of the alternatives, but . . . only correlated with the desired outcome") and objectives ("directly connected with the desired outcomes, but for which the estimation will be in general fuzzy").

Not only is the identification of the criteria set a difficult task – often guided by no more than educated intuition – it is something that deserves repeated revision during the planning process so as to reflect as many and as much of the values of affected participants as possible.

By and large, the most uncontroversial part of MCS procedures is the analysis of alternatives: this is the only area where relatively nonarbitrary procedures are available, at least at the initial stage. For each $k_i \in K$, let $\mathbf{f}_i : A \to \mathbf{Z}^+$, where $\mathbf{Z}^+$ is the set of positive integers, generate the weak order induced by k_i.[30] Let $w_{im} = \mathrm{f}_i(\alpha_m)$; thus the w_{im} form a matrix (which is infinite if A is infinite). Thus w_{im} is the value of the mth alternative according to the ith criterion. Each row of this matrix corresponds to a different criterion: thus incommensurable criteria are never being reduced to the same scale. (Each column of the matrix corresponds to a different alternative.)

An alternative $\alpha_m \in A$ is *dominated* by another alternative $\alpha_n \in A$ (with respect to K)[31] if and only if:

$$\exists p, k_p \in K, \text{ such that } w_{pn} < w_{pm}$$

and

$$\forall q, k_q \in K, w_{qn} \leq w_{qm}.$$

This means that α_n is strictly better than α_m by at least one criterion, $k_p \in K$, and is not worse than α_m by any criterion $k_i \in K$ (obviously including k_p). An alternative $\alpha_i \in A$ is *non-dominated* if and only if it is not dominated by any other element of A. The non-dominated subset of A, $A_N \subset A$, is the set of all non-dominated elements of $A : A_N = \{\alpha_i \in A : \alpha_i \textit{ is non-dominated}\}$.[32]

Given A, MCS strategies should begin by identifying A_N. The justification for this is fairly straightforward and, most importantly, does not involve any arbitrary assumption. Consider any dominated element of A, say α_m. Then there is at least one element of A_N, say α_n, such that α_n is better than α_m by at least one criterion and no worse by any criterion. So long as we have α_n, it seems fully reasonable to drop α_m from further consideration. Typically there is more than one non-dominated alternative. As the next section will show using an example, when there are several such alternatives, but not so many that they are intractable for further analysis, this multiplicity of equally

[30] Obviously, for each i there can be infinitely many such $\mathbf{f}_i$, because all that has to be maintained is the ordinality of the values of the functions.

[31] For the sake of expositional simplicity, this qualification will henceforth be dropped.

[32] In economic contexts, it is also called the "Pareto optimal set."

acceptable solutions is a virtue: all of these alternatives can be presented to decision makers for further refinement using other criteria, including the intangible ones mentioned before.

Unfortunately, in many cases the cardinality of the non-dominated set may be high enough to make it practically intractable for further analysis by decision makers.[33] In the general case, beyond finding the non-dominated set, no further refinement of the alternatives is possible without introducing partly arbitrary assumptions. This is where even more techniques that do not belong to conventional economics come into play. As noted before, the response of conventional economics is to attempt to find a single utility function incorporating all the criteria. The easiest way to find such a function is to represent the utility function as a linear sum of the different criteria.[34] However, more sophisticated strategies involving the elicitation of criterion rankings, along with alternative rankings, are also available.[35] While everything that was said above about non-dominated alternatives is consistent with conventional economics, and while utility theory also prioritizes non-dominated alternatives above dominated ones,[36] the strategy of first finding the non-dominated alternatives departs from the usual procedure of conventional economics. However, it is at the stage of choosing between non-dominated alternatives that unconventional techniques become most important. These range from treating the problem as one of finding optima in an abstract space (with no concern for finding an underlying utility function) to ranking preferences by a variety of processes. In the context of biodiversity conservation, perhaps the most popular strategy for developing such a ranking has been the analytic hierarchy process (AHP).[37] However, unlike the computation of non-dominated alternatives, all these strategies involve ad hoc assumptions. They will not be discussed further

[33] In general, the number of non-dominated solutions increases with the number of criteria – for a more detailed discussion, see Sarkar and Garson (2004).

[34] In the context of biodiversity conservation, significant attempts to do so include that of Faith (1995). Roughly, Faith incorporates biodiversity and economic considerations as "costs," with a parameter determining the relative importance of each. If a result is consistent with a wide variation in the value of this parameter, then it is taken to be reliable.

[35] The extensive theory developed to this end is called multiple attribute utility theory (MAUT) (Keeney and Raiffa 1993) when uncertainty is taken into account; when uncertainty is not taken into account, the result is multiple attribute value theory (MAVT) (Dyer and Sarin 1979). These are being regarded as part of conventional economics here because they involve the elicitation of utilities (or the corresponding "values," in the absence of uncertainty).

[36] Note that the non-dominated sets correspond to the indifference curves of conventional economics.

[37] For an account of the analytic hierarchy process (AHP), see, for instance, Saaty (1980). For a review of applications to biodiversity, see Moffett et al. (2005).

here – rather, the next section will turn to the application of the computation of non-dominated alternative sets in the context of place prioritization for biodiversity conservation.

7.3. THE CONTEXT OF BIODIVERSITY

When attempting to produce a final prioritization of places in the context of biodiversity conservation, whether further analysis, beyond the identification of the non-dominated alternative set, is necessary will critically depend on the size of the non-dominated set, A_N. At one extreme, in the most trivial (and practically irrelevant) case, if A_N contains exactly one element, α, then no further analysis is necessary: α is the only relevant alternative. At the other extreme A_N may be the same as A. The important point to note is that, as mentioned at the end of the last section, the set of "best" solutions cannot in general be further refined using rational considerations alone. In this sense, arbitrary (and, in that sense, ad hoc) assumptions of varying degrees of plausibility – often resulting in variations of the classical economists' attempts to reduce all values to a single scale – are necessary in order to go any further. In the specific context of biodiversity conservation, it may be possible that purely rational methods can be devised that will allow such further refinement. This remains an open question. At present, all that can be said is that there is no purely rational analytic procedure for further refinement of the nondominated solution. The identification of nondominated solutions is the place where rational or "systematic" conservation planning ends and political and other considerations, including the use of intangible values, should enter.

The size of A_N depends on whether terminal stage or iterative procedures (using the terminology of § 7.2) are used. In almost all situations, the former generate much smaller alternative sets than the latter. Typically, terminal stage procedures are used when biodiversity is given qualitatively different preferential treatment over other values, that is, when it is privileged. A set of different preliminary lists of prioritized places are created, for instance, using the heuristic procedures described in Chapter 6 (§ 6.2), each of which meets whatever biodiversity target has been set. These are the *feasible* solutions. The other criteria are now used to find non-dominated solutions from the set of feasible solutions. In this procedure, biodiversity representation is never compromised. If the true surrogates for biodiversity that are chosen are taxa that are at risk of extinction, this is the method that will appeal to most biodiversity conservationists. For this reason, this chapter started with such a scenario.[38] In such a circumstance, because the feasible alternative solution set is usually small, there will generally be very few non-dominated solutions. These can be turned over to the political process, where other considerations (including intangible values) can be brought in. Since there are few alternatives, and since the alternatives are final proposed networks of conservation areas, what is left for policy makers to consider should not appear to be too onerous. The experts will have provided exactly the kind of input that most policy makers presumably desire: a short list of final alternatives that are ranked equally by rational or scientific considerations.

If, however, biodiversity conservation is viewed as a value on a par with other values, such as economic and social costs, then an iterative procedure may be used to find non-dominated solutions at each iterative stage. In this scheme, place prioritization must itself take these other features into account, and the exact or heuristic procedures of the sort described in Chapter 6 (§ 6.2) must be modified accordingly. One typical problem with such procedures is that of finding large non-dominated sets at each stage of the iteration. With no rational procedure for paring these, choosing one of the places arbitrarily is as good an option as any other. This procedure will not be explored any further in this book, though it has its advocates.[39]

An example from Texas will clarify the methods that have just been described. In this example biodiversity was privileged over other values in the sense of the last section: satisfaction of the biodiversity targets was first ensured before other values were considered. Thus, this is a terminal state procedure. Using the same GAP analysis data set that was described in Chapter 6 (§ 6.2) in the context of discussing place prioritization, 100 attempts were made to generate different prioritized lists of hexagons, with a target of 10 representations of each vertebrate species. Technically, since the biodiversity content of the hexagons was not probed for viability, this amounts to multiple criteria synchronization on the basis of biodiversity content, not value. Because of the difficulty of carrying out adequate viability analyses in most places, this will unfortunately be typical in contexts of policy formulation and analysis.

There were thirty-two distinct feasible solutions, each of which satisfied the biodiversity conservation target. Two criteria other than biodiversity were then considered: economic cost and social cost. Economic cost was

[38] For a different anthropocentric perspective, which also gives a qualitatively different status to biodiversity and is, therefore, consistent with the procedure outlined here, see Wood (2000).

[39] See, in particular, Faith (1995).

estimated by the total area that would be put under some sort of conservation plan, that is, the total area of the selected hexagons. This estimate assumes that all areas have the same cost.[40] Social cost was estimated by the human population of each hexagon, the assumption being that the number of people affected by a plan is as good a measure of social cost as any other that is likely to be available.[41] However, in general, in the context of practical policy formulation, uncontroversial measures of social cost are unavailable.[42]

Figure 7.3.1 shows the thirty-two possible solutions plotted against social and economic cost. The two non-dominated solutions are shown as circles; the other feasible solutions are shown as squares. Figure 7.3.2 shows the locations of the hexagons selected by the two solutions. The distance between solutions may be estimated by the ratio of the number of unshared cells to the total number of cells in the two solutions (considered separately). This distance measure will vary between 0 and 1. It is an appropriate measure: when two solutions are identical, the distance is equal to 0; when they are completely disparate, it is equal to 1; and, finally, the distance increases linearly with the number of unshared cells. The distance between the two solutions depicted in Figure 7.3.2 is 0.534. This shows that the disparity between the two solutions is quite high. Were this exercise intended to be a policy recommendation, both non-dominated solutions would be presented to policy makers for further consideration. It is likely but not certain that the shared cells have greater biodiversity value than the unshared.

In Figure 7.3.1, the two non-dominated solutions are easily visually identified: they are those two solutions that have no other solution in their lower left-hand quadrant. This means that no solution is better than either of the non-dominated solutions along both axes, that is, according to both of the criteria. This procedure of identifying non-dominated solutions can

[40] This is undoubtedly a false assumption; therefore, the approximation involved is unjustified. It is being used here because actual costs are unavailable. However, recall from Chapter 6 (§ 6.2) that the analysis of data from Texas is intended merely as an illustration of the methodology and not as an actual policy recommendation. Consequently, the approximation does no harm. In general, in almost all contexts where place prioritization is carried out with the aim of policy formulation, these costs will be available.

[41] Once again, it is easy enough to criticize the approximation involved. For instance, the social cost of affecting a poor individual by some action should be regarded as being higher than that of affecting a rich individual in the same way; the latter will have more options available than the former for ameliorating the extent of difficulties that may be produced by the performance of that action. Social costs vary with the nature of the habitat, the nature of the individuals involved, and so on.

[42] For an introduction, see Barrow (2000).

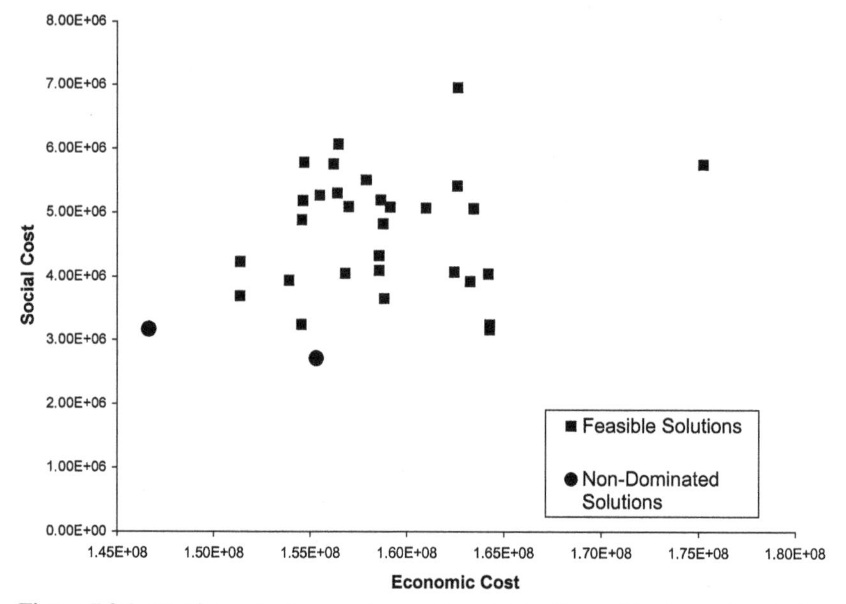

Figure 7.3.1. Multiple criterion synchronization plot for Texas. The figure depicts the thirty-two distinct feasible solutions (those that met the biodiversity target) for Texas. Economic cost is estimated along the x-axis as the area of a selected cell; the units are in acres. (For instance, 1.40E+08 should be interpreted as 1.40×10^8 acres.) Social cost is estimated along the y-axis as the human population of a selected cell.

be generalized to any number of dimensions, that is, to any number of incommensurable criteria, though, obviously, straightforward visualization is impossible in dimensions greater than three. In order to apply this procedure, all that is required is that when each criterion allows the placement of alternatives on its ordinal scale, the imposed order gives lower ranks based on how desirable an outcome is. Note, finally, that the lengths along the axes do not necessarily convey meaningful information: this procedure uses only ordinal information.

7.4. COPING WITH UNCERTAINTY

Recall the discussion of extinction in Chapter 5 (§ 5.3). The estimation of extinction rates was so fraught with uncertainly that it was not even possible to be certain whether there is an ongoing biodiversity "crisis" even after a convention-based definition of crisis was adopted. This predicament raises

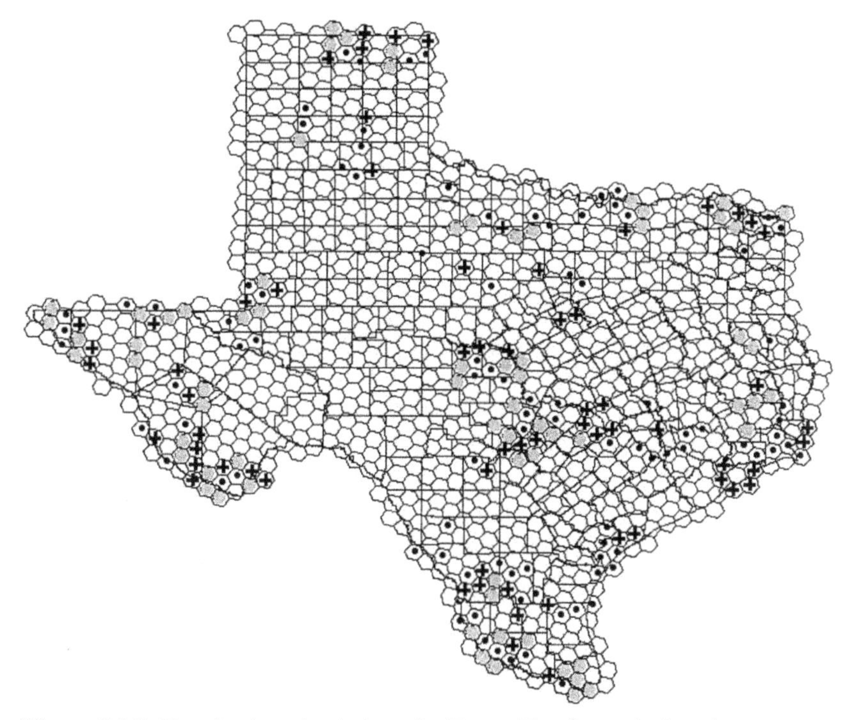

Figure 7.3.2. Non-dominated solutions for Texas. The figure depicts the two non-dominated solutions. Gray cells are shared by both non-dominated solutions; cells with black dots are in the first but not the second solution; cells with crosses that are in the second but not the first solution. For further discussion, see the text.

an interesting but difficult question. Does prudence decree that, faced with uncertainty, we declare a crisis because that will very likely lead to conservationist measures, whereas no action may lead to permanent extinctions? This is one, rather controversial use of the "no harm" principle (sometimes also called a "precautionary" principle).[43] There is much to be said for this principle, even though it is essentially conservative and usually dictates against proactive policy adoption. Nevertheless, its use in this context is controversial, to say the least, because a decision to assign a place to biodiversity conservation may conflict with other uses of that place. These other uses may reflect criteria other than biodiversity conservation that may be critically important to individuals, and to prevent satisfaction of those criteria may therefore constitute some harm. This is one reason for hesitation about declaring an ongoing biodiversity crisis. Moreover, as noted in

[43] When in doubt over a set of alternative actions, choose the one that is least likely to do any harm.

Chapter 5 (§ 5.3), it is far from clear that prudence decrees talk of "crisis"; rather, it may decree avoidance of any action that may adversely affect the credibility of biodiversity conservationists among those who frame public policy.

The critical point is that biodiversity conservation requires action and does so in the face of largely irremediable uncertainty. Coping with uncertainty raises well-known philosophical problems that have received considerable attention but little resolution. That uncertainty must be addressed has long been explicitly recognized in policy formulation based on the social sciences. There have been many attempts to incorporate uncertainty, explicitly and quantitatively, in protocols for decision making. Thus far, although the relevance of uncertainty has always been explicitly recognized in the context of biodiversity conservation, it has rarely been explicitly formally incorporated into conservation planning. Incorporating considerations of uncertainty quantitatively is the most important task for the conservation biology of the future. What follows in the rest of this chapter should be regarded as no more than setting the stage for the work that remains to be done.

Moving beyond comparatively straightforward cases of uncertainty, such as that of estimated extinction rates, there are at least four different ways in which uncertainty enters into biodiversity conservation planning:[44]

(i) *Structural uncertainty*: alternate models of the same biological situation make radically different predictions. This may happen because biological mechanisms are not well understood, or because approximations that have to be introduced in order to construct a tractable model introduce uncertainties that cannot be controlled. An example of this sort was seen in the last chapter (§ 6.4) when two slightly different models of a population experiencing demographic and environmental stochasticity made widely divergent predictions of the expected time to the extinction of the population. Insofar as prediction of such parameters is concerned, other than admonishing decision makers to understand biological mechanisms better and to analyze the formal properties of models more adequately, there is little that can be suggested to alleviate this problem. However, as noted in Chapter 6 (§ 6.4), some robust conclusions, such as whether a particular population is more likely to become extinct than another under all models,

[44] The terminology is that of Williams (1997), though the description of the categories departs significantly from that treatment.

are still useful. Bayesian statistics (see § 7.6) also sometimes allow the combination of results from different models, thus reducing structural uncertainty. Note that this uncertainty is independent of uncertainties due to environmental variations or in the estimation of parameters in the field.

(ii) *Environmental uncertainty*: that environmental features may vary in random ways is a feature of every biological system. This type of uncertainty can be partially modeled using the various models of demographic, environmental, and random catastrophic stochasticity that were briefly introduced in Chapter 3 (§ 3.2) and also discussed in the last chapter (§ 6.4). However, it is far from clear that the models so far constructed are empirically adequate. Once again, much remains to be done. Moreover, the structural uncertainties in the models discussed earlier remain unresolved.

(iii) *Partial observability*: as mentioned in Chapter 3 (§ 3.2), a problem that bedevils theoretical ecology is that even the simplest of parameters, such as the intrinsic growth rate of a population or the carrying capacity of a habitat, are almost intractably difficult to estimate accurately in the field. The uncertainty in parameter estimation can potentially be amplified when these parameters are used in models that are themselves highly sensitive to the value of parameters (point [i] above). In the context of biodiversity conservation, an added uncertainty arises when biogeographical data compiled from heterogeneous sources have to be used simultaneously. For instance, all place prioritization procedures utilize data about the distribution of surrogates. These data are generally treated as presence-absence data. But, as noted in Chapter 6 (§ 6.2), some of these data are presence-only but nevertheless must be used for planning purposes, because no other data are available. How presence-absence data should be compounded with presence-only data remains unclear; no convincing procedure has been produced to date. If the surrogate distribution data used are from the predictions of models, then the presence of a surrogate at a given place can be predicted only as a specified probability. How these probabilities may be incorporated into a place prioritization procedure will be indicated later (in § 7.5). For once, the problem is not particularly difficult. How some of the other problems may be partially mitigated, though not solved, will also be discussed later. Bayesian estimation may provide a way out of some of these difficulties (§ 7.6).

(iv) *Partial controllability*: the effects of management plans are not only hard to predict, they are also hard to control. This situation adds yet another level of uncertainty to the process of adaptive management. Once again, a Bayesian decision-making procedure promises some alleviation of the associated problems, as will be very briefly indicated later (§ 7.6).

7.5. PROBABILISTIC PLACE PRIORITIZATION

Some of the problems arising from uncertainty are not difficult to mitigate, at least partially, even though they have rarely been broached in practice. For instance, the problem of having only probabilistic expectations for the presence of surrogates at each place during a place prioritization process is not difficult to "solve," at least in one sense.[45] The relevant sense of "solve" is that probabilities of presence can be incorporated into a plausible algorithm for prioritizing places. This can be achieved by interpreting the probabilities as expectations.[46] Let there be n places in a set Σ of places and m surrogates. Let p_{ij} be the probabilistic expectation of the ith surrogate at the jth place. These p_{ij} are typically the numbers predicted by the models used to obtain the distributions of surrogates for use by place prioritization procedures. If $p_{ij} = 1$, then the ith surrogate is present with certainty in the jth place. Such an assignment can be made (or at least p_{ij} can be assigned a value very close to 1) from reliable positive reports in both presence-only and presence-absence record sets.

If $p_{ij} = 0$, then the ith surrogate is absent from the jth place. This assignment can be inferred only if one has very reliable presence-absence record sets or significant knowledge of the ecology of the surrogate. If what is available is only a presence-only record set, the absence of a positive report cannot be used to infer that $p_{ij} = 0$. It is an open question what value should be attributed to the p_{ij}. A traditional principle of indifference may suggest that probabilities should be equally distributed between two possibilities: (i) the surrogate is present but not observed, and (ii) it is absent. In almost all circumstances, such a distribution of probabilities would give too high a value for the p_{ij}. The value that should be used is likely to be

[45] For the place prioritization problem, one of the few previous attempts that does not make ad hoc assumptions is that of Polasky et al. (2000). For a systematic discussion of the issues treated in this section and a review of the literature, see Sarkar et al. (2004).

[46] Expectations provide an alternative framework for quantifying uncertainties than the usual framework of probabilities. For more on this poorly explored perspective, see Whittle (2000). Its application to place prioritization is discussed in Sarkar et al. (2004).

highly context-dependent: it varies with every factor that influences whether a surveyor would record a surrogate that is not being actively searched for; and it also depends on whether this record would survive subsequent data treatment, especially if information that is perceived to be irrelevant is excised. It is quite likely that the p_{ij} are indeterminate in this case. Opting for $p_{ij} = 0$ would then be a conservative "safety first" or "precautionary" choice: if the ith surrogate is still protected up to the desired target in a set of selected places, the uncertainty in the p_{ij} ensures only that it may have exceeded the target. What is lost is efficiency in the selection of the prioritized set of places.[47]

Given the p_{ij}, $s_i = \sum_{j=1}^{n} p_{ij}$ is the expected number of places in which the ith surrogate will be present in the entire data set. Similarly, $b_j = \sum_{i=1}^{m} p_{ij}$ is the expected number of surrogates present in the jth place. Let $\Pi = \{\pi_1, \pi_2, \pi_3, \ldots, \pi_l\}$ be a set of l places prioritized for protection. Then $s_j^{\Pi} = \sum_{\pi_j \in \Pi} p_{ij}$ is the expected number of places in Π at which the ith surrogate is present. The first canonical form of the place prioritization problem (from Chapter 6, § 6.2) now becomes that of finding the set Π of lowest cardinality for which all the s_i^{Π} meet their required targets. The second canonical form of the place prioritization problem now becomes that of finding, for a given l, the set of places Π of cardinality l that maximizes the number of surrogates for which the s_i^{Π} meets the required target.

It is now possible to incorporate the expected presence of surrogates at places into a heuristic rarity-complementarity–based algorithm. Targets must now be interpreted as the values that the s_i^{Π} must achieve for each surrogate. As before, the algorithm can be initialized by richness (that is, the place with the highest b_j), by rarity (see below), or by using a list of existing selected places. The rarest surrogate is the one with the lowest s_j; the place with the best rarity rank is the one for which the rarest surrogate has the highest probability of presence. The complementarity value for a new potential addition to Π will be the sum $b_j = \sum_{i=1}^{m'} p_{ij}$, where m' indicates that the summation is carried out only for those surrogates for which the target has not yet been met. How well this algorithm performs remains a topic of ongoing research.[48]

[47] For a discussion of situations analogous to this in terms of (frequentist) statistical type I and type II errors, see Shrader-Frechette (1990).

[48] Polasky and colleagues (2000) have explored a related algorithm that does not use rarity with inconclusive results; Sarkar and colleagues (2004) found more promising results using complementarity and, unlike the situation with binary presence/absence data, less promising results using rarity and complementarity simultaneously.

7.6. A BAYESIAN FUTURE?

The most important methodology that is available for incorporating considerations of uncertainty into biodiversity conservation planning is the use of Bayesian statistics. This approach allows the systematic incorporation of uncertainty at every level – from parameter estimation, through the results from a model, to the context of making decisions. These claims about the power of Bayesian statistics are routinely questioned in philosophy, in statistics, and in the empirical sciences.[49] For those who disagree about the value, or even the validity, of Bayesian statistics, the remaining portion of this chapter will be controversial. There is insufficient space here for an adequate defense of Bayesianism, and none will be attempted.[50] The stance adopted here is that Bayesian statistics provides more hope for coping with uncertainty in conservation biology than other competing statistical methodologies (which may well be more appropriate in other contexts). If a positive agenda for biodiversity conservation is the goal, then critics, beyond merely criticizing, should provide alternative analyses that fare better than the ones produced by Bayesians.

There are two main methodologies within statistics: so-called frequentist statistics and Bayesian statistics. The former is so much more popular than the latter that it is usually simply denoted "statistics"; here it will be referred to as "conventional statistics."[51] Consider the problem of partial observability, which results in uncertainties in the estimation of parameters in the field. Conventional statistics can be used to estimate the value of a parameter as a single number (a "point-estimate") with an associated confidence interval. It can also be used to calculate a probability value, the p-value, of a precise hypothesis about the reliability of the estimate. It is important to be clear what this hypothesis is: the p-value cannot be interpreted as the probability that the parameter has the value estimated for it; rather, given that value for the parameter, the p-value is the probability of not collecting the observed data were the population to be repeatedly sampled. Similarly, a confidence

[49] In the context of ecology, see, for instance, the criticism of Dennis (1996). See, however, Ellison (1996) for a contrary perspective.

[50] Suffice it to say that it is being assumed here that such a defense is more than possible. See Robert (2001).

[51] For the customary terminology, see, for instance, Wade (2000). The choice of the terminology adopted here is motivated by the philosophical point that the type of statistics we use is independent of the interpretation of probability (logical, subjectivist/personalist, propensity, or frequency) we choose. Moreover, not all frequentists (going back to Fisher) accept conventional statistics, including its tests of significance.

interval – say, of 95 percent – can be interpreted as saying only that, were the data sets repeatedly sampled, the true value would lie within the calculated interval 95 percent of the time. It does not say that there is a 95 percent probability that the true value lies in the calculated interval, which, in any case, is not a probability distribution.

In sharp contrast, Bayesian estimation results in a probability distribution for the value of the parameter. Bayesian estimation of a parameter depends on a basic rule (Bayes' rule) that connects a posterior probability distribution for a parameter ("posterior" in the sense of epistemologically following data collection) and a prior probability distribution for that parameter using the likelihood of the data if the prior distribution were the correct one:

$$\text{posterior distribution } (\pi) \propto \text{prior distribution } (\pi) \times \text{likehood } (\delta | \text{ prior distribution}),$$

where π is the parameter being estimated and δ is the data that were collected. Both the prior and the posterior distributions give the probability that the parameter has a particular value.[52] Essentially, this is an updating rule that can be used iteratively, with the posterior distribution from one analysis becoming the prior distribution for the next. If an interval captures α percent of the probability distribution (as the area under the curve), then there is an α percent probability that the value of the parameter lies within that interval. Bayesian estimation is straightforward and transparent.

Bayes' rule is a generalization of a theorem from elementary probability theory after that theorem is given a particular epistemological interpretation. Let α and β be events. Let $\mathbf{p}(\alpha)$ be the probability of α and $\mathbf{p}(\beta)$ be the probability of β. Let $\mathbf{p}(\alpha \mid \beta)$ be the probability of α given β; by an axiom of the probability calculus (in its usual form), $\mathbf{p}(\alpha \mid \beta) = \frac{\mathbf{p}(\alpha \wedge \beta)}{\mathbf{p}(\beta)}$. Then it is an elementary theorem of the probability calculus that

$$\mathbf{p}(\alpha \mid \beta) = \frac{\mathbf{p}(\beta \mid \alpha)\mathbf{p}(\alpha)}{\mathbf{p}(\beta)}.$$

This theorem has considerable epistemological significance if α is interpreted as the event of a hypothesis being true and β is interpreted as the

[52] Point estimates can also be obtained from Bayesian statistics by incorporating a loss function with the posterior distribution and subsequently minimizing the posterior expectation of the loss function. Credence intervals (analogous to conventional confidence intervals) can also be generated. However, from a Bayesian point of view, point estimates and credence intervals are only convenient ways of summarizing the full posterior distribution function and have no further significance.

event of finding the evidence. Then $\mathbf{P}(\beta \mid \alpha)$ is a likelihood and $\mathbf{p}(\alpha)$ and $\mathbf{P}(\alpha \mid \beta)$ are the prior and posterior probabilities of the hypothesis, α. Under this interpretation, this equation is known as Bayes' theorem and provides the basis for Bayesian statistics. If Bayes' rule, as given in the last paragraph, were converted into an equation through the introduction of a normalization factor as the constant of proportionality, that equation would simply generalize the one given in this paragraph, replacing probabilities for single events with probability distributions.

The main problem usually attributed to the use of Bayesian statistics is that a prior probability distribution is necessary for the use of Bayes' rule. The usual objection is that choosing a prior distribution for the value of a parameter before measurement (or, alternatively, attributing a prior probability to a hypothesis before confrontation with data) is arbitrary. This is an important objection. Nevertheless, Bayesians have devised strategies to mitigate its force. "Objective" Bayesians generally prefer uninformative priors: for instance, in the absence of any information whatsoever regarding the value of a parameter, a principle of indifference would suggest that all possible values be ascribed the same prior probability.[53] "Subjective" Bayesians would prefer to use whatever knowledge may be available about the parameter to inform the selection of the prior distribution.

But it is possible to make a virtue of the requirement that a prior probability distribution must be assumed in the use of Bayes' rule: it allows the incorporation of biological knowledge into the calculation of the uncertainty of a result – including, especially, the results of previous experiments or sets of observations. Suppose, for instance, the size of a population has to be estimated, as is required in almost all models of population viability analysis.[54] Suppose that several independent surveys, each with an associated uncertainty, are carried out. Conventional statistics provides no method to combine the results of these surveys in order to reduce the uncertainty through a composite analysis. Bayesian analysis is ideally suited to that task. Any of the surveys can be used first (that is, the order does not matter). Starting with some prior – objective Bayesians would use an uninformative prior, subjective ones would try to incorporate any hint of biological

[53] Naive versions of the principle of indifference are known to lead to paradoxes – see Horwich and Urbach (1993), pp. 59–62. However, more sophisticated ways of incorporating as little information into the priors as possible – for instance, the maximum entropy method – have been devised; see Robert (2001), pp. 109–111. The maximum entropy method minimizes the Shannon information incorporated into the prior probability distribution.

[54] Wade (2000) carries out such an analysis for a hypothetical population, illustrating all the methods that have to be used.

knowledge they may have – Bayes' rule is used to generate the posterior probability distribution. Then this posterior distribution can be used as the prior to analyze the result of another survey. The process can be iterated until all surveys have been used. Typically, the result will be an increasingly sharper distribution for the size of the population and a concomitant decrease in uncertainty.

Perhaps the greatest utility of Bayesian methods in the context of systematic biodiversity conservation planning is that it provides a rational framework for making decisions in the presence of uncertainty when there are many options to be considered. In sharp contrast, conventional statistics does not allow the simultaneous consideration of more than two options (if that): in the context of formal decision theory, not being a Bayesian is usually not a rational decision to make. Bayesian decision theory requires the introduction of a loss function that formalizes the loss connected with each decision from a set of possible alternatives. Decisions in the alternative set have probabilities associated with them and can be of various types. Generally, they arise from a posterior probability distribution obtained through the use of Bayes' rule: for instance, a parameter estimate can be considered to be a decision to accept that value for the parameter; the associated probability is then obtained from the posterior probability distribution for that parameter. This probability is now multiplied by the loss function, and the product is minimized for the entire set of alternatives.

This process can be illustrated using an example that will also show how Bayesian estimation is naturally incorporated into a decision-making process.[55] The problem is the classification with respect to risk of extinction of the Spectacled Eider, *Somateria fischeri*.[56] These are large-bodied sea ducks that nest along the Arctic coast of Siberia and the Chukchi and Bering Sea coasts of Alaska. In 1993, the United States Fish and Wildlife Service listed the Spectacled Eider as threatened; what was at stake was whether its classification should be changed to "endangered." The decision process of the Spectacled Eider recovery team involved four stages:

(i) Choice of a *data set*: the data set used consisted of abundance index estimates. These were available only for the Yukon – Kuskokwim delta, that is, the southern Alaskan meta-population. Data were available from three independent surveys.

[55] For a different method of deploying the Bayesian framework in population viability analysis, see Goodman (2002).

[56] See Taylor et al. (1996) for the report on which the discussion in the text is based.

(ii) Choice of *classification* criteria: whether a species is "endangered" or "threatened" depends on how those terms are defined, which is not determined by scientific criteria alone. Based on simulations including environmental stochasticity, the Spectacled Eider recovery team deemed a population to be endangered provided it was declining by ≥ 5 percent per year.

(iii) Choice of *decision* criteria: in Bayesian analysis, this involves the choice of a "loss function" that attributes an explicit loss to a decision, that is, to a (mis)classification. Two types of loss functions were used: (a) a simple loss function, which assigned a loss of one to an incorrect decision not to classify the Spectacled Eider as "endangered" and a loss of zero to other decisions; and (b) a more complex continuous loss function, which tried to quantify the risk of misclassification as a function of the rate of decline.

(iv) *Analysis*: first, a Bayesian trend analysis was performed.[57] The use of Bayesian methods allowed the simultaneous use of data from all three surveys. The prior probability distribution of the population size was assumed to be a normal distribution; a uniform prior probability distribution was used for all the other parameters so as to reflect no prior knowledge about their values. The region for which the annual population growth rate was ≤ -0.05 encompassed 99.86 percent of the posterior distribution. Both loss functions resulted in the population being classified as endangered in all but one analysis (with the data sets taken singly or in all possible combinations).

Conventional statistics could not have provided such an unequivocal rationale for classifying the Alaskan Spectacled Eider population as endangered.

There are at least seven advantages to the use of Bayesian over conventional statistics in the context of biodiversity conservation planning:[58]

(i) When parameters are estimated, Bayesian analyses allow inclusion of the uncertainty of the estimate of each parameter (through a "hierarchical" modeling process). Thus, for example, a Bayesian analysis can produce a statement of the uncertainty of an estimate of a probability of extinction within a specified time period. Conventional statistics cannot directly produce such statements.

[57] Wade (1994) describes the method, which is an analog of nonlinear regression in conventional statistics.

[58] This list is partly based on Wade (2000).

(ii) The Bayesian posterior probability distribution for a parameter estimate directly gives the probability that the parameter has a particular value. There is no corresponding interpretation of a conventional statistical estimate.

(iii) Bayesian analyses can incorporate whatever biological knowledge that is available into the statistical procedure, in particular, through the introduction of prior probability distributions. There is no similar resource in conventional statistics.

(iv) Uncertainty from parameters that are known to be important but the values of which are unknown can be incorporated into the analyses using what are called "hierarchical" Bayesian models.

(v) Uncertainty can be decreased in a transparent way, for instance, by sequentially refining priors by collecting more data through additional surveys and experiments.

(vi) Bayesian decision theory coheres seamlessly with other Bayesian techniques to provide a formal framework for decision making that also incorporates uncertainty. What was said in (v) about decreasing uncertainty carries over to this case as well.

(vii) Finally, there is also a technique – which merits much more philosophical scrutiny than what has so far been afforded to it – that allows uncertainties about models to be incorporated into the analyses by combining predictions from different models. This involves the use of a parameter called the "Bayes factor." So far, it does not seem to have been used in the context of biodiversity conservation.[59]

(It should also not go unnoticed that conventional [maximum likelihood] parameter estimation can be subsumed under Bayesian methods by choosing a uniform prior probability distribution and choosing the mode of the posterior probability distribution as the point estimate.[60] Thus even those who have philosophical objections to Bayesian methods should recognize their greater generality.) The use of Bayesian methods in the context of biodiversity conservation is only beginning. It has a good chance of transforming the practice of conservation biology.

[59] See Kass and Raftery (1995).

[60] See, for instance, Asteris and Sarkar (1996) and the references therein.

8

In Conclusion

Issues for the Future

This book will not have a traditional conclusion summarizing whatever insights that are presumed to have been provided by the earlier chapters. Rather, this concluding chapter will list and emphasize issues that have been treated only inadequately in the preceding chapters. Professional philosophical interest in biodiversity conservation is so recent that much more work remains to be done, compared to what has so far been accomplished. Most questions have been answered only incompletely, if at all: they remain largely open. In one sense, the conclusion of the last sentence is always true in philosophy construed as an exploration of perennial questions rather than as an attempt to provide definitive answers to them. However, that is not the sense in which the questions discussed here remain open. As also noted in the Preface, professional philosophical work in this area is of very recent vintage. Consequently, whereas almost all the issues explored in this book have received some attention from professional philosophers, except for most of those connected to environmental ethics none have received the detailed attention they deserve. More importantly, they have not received systematic attention jointly. This is a first attempt to bring together the ethical and epistemological issues that emerge from our concern for the environment, especially the depletion of biodiversity, and our attempts to respond to them.

The next section will list four issues surrounding the axiological (normative ethical and, on occasion, aesthetic) analyses of Chapters 2–4 that merit further exploration; the section that follows will list six of the corresponding epistemological issues arising from the more scientific discussions of Chapters 5–7.

8.1. THE VALUE OF BIODIVERSITY

Suppose, for the moment, that attributions of transformative value make sense and that the arguments given in Chapter 4 are at least plausible – problems will be duly noted at the end of this section. Even with this assumption, at least four partially unresolved questions arise from the normative ethical and aesthetic discussions earlier in the book:

(i) Does the appeal to transformative value capture all that we find valuable in biodiversity? Entities with transformative values, according to the account developed in this book, acquire that value by their potential to transform the demand values of individuals. But there are many features of biodiversity that are directly attributed to cultural, rather than individual, value, and many strategies for conservation are based on a recognition of that value. Among the First Nations of North America, many animals have totemic value and deserve protection for that reason. The Bengal tiger (*Panthera tigris tigris*) is so firmly embedded in the myths and folklore of Bengalis (of Bangladesh and India) that its extinction would result in a serious cultural loss. These cultural values can obviously be used to promote biodiversity conservation and have sometimes contributed toward that end.[1] In many cases, traditional practices based on cultural values have led to the protection of many species. Can such cultural values provide an adequate basis for the conservation of biodiversity in general? It does not appear likely. Trivially, there is the problem that not all cultures have had such ongoing traditions, at least over the last few centuries. Such cultures include, most importantly, the Northern cultures that dominate the world economically and politically. But, more importantly, as Gary Nabhan has pointed out, concern for and adequate management of individual features of biodiversity (for instance, species) does not automatically translate into a concern for general biodiversity.[2] No traditional culture is known to have conserved biodiversity in general; obviously, this claim is open to empirical refutation.

[1] Gadgil, in particular, has very persuasively argued for the role of sacred groves in preserving the few remaining patches of primary tropical wet evergreen forests in the Western Ghats of India (see, for example, Gadgil and Berkes 1991 and Gadgil and Guha 1992).

[2] Nabhan (1995) distinguishes between resource management, for which there is ample evidence, and conservation of all of biodiversity, for which there appears to be very little.

Can such cultural values be reduced to individual ones and, especially, to individual transformative values?[3] Once again, what is at stake is methodological individualism (in this case, in the social sciences). If such a reduction is not possible, the account of transformative value in Chapter 4 cannot be directly used to capture cultural values. However, it is probably not incoherent to think of entire cultures being transformed by some experience beyond the transformations of individuals who comprise that culture. Even if the relevant reductions can be carried out, there is a deeply troubling feature of this move. Prior to the putative reduction, there is no single "cultural value"; rather, there are different values shared to different extents by members of the same culture. Treating these different values independently was the whole point of the multiple criteria synchronization (MCS) procedure. Yet founding all of them on transformative values amounts to reducing them to the same scale for foundational, if not mensurative, purposes. For the positions explicitly defended in this book, this may not pose a problem, because transformative values were left unquantified. But there was no argument given to indicate that they cannot or should not be quantified. The question of quantifiability was ignored because, logically, no answer to it was required to further the arguments of this book.

(ii) Is there much more to be learned from the similarities between biodiversity and objects of aesthetic interest than what has been broached in this book? This question bears much more exploration than it can be afforded here. In Chapter 4 (§ 4.4), it was suggested that works of art (and, also, wilderness) may acquire their value by their ability to transform the demand values of individuals. Such transformative ability was taken to be enough to generate obligations in us to conserve biodiversity. In the aesthetic analogy, the same argument generates an obligation to protect objects deemed to be works of art, whether or not we appreciate their aesthetic qualities ourselves. Similarly, it becomes obligatory to protect many other cultural artifacts, including those of historical, religious, and other cultural significance. What is interesting, and indirectly provides yet another argument for the use of transformative value as a basis for conserving biodiversity, is that our common practice seems to recognize these obligations in exactly this way: in the United States, just as there are National Parks to protect

[3] "Reduced to," in this context, means "explained on the basis of" – see Sarkar (1998a) for a development of this (philosophical) view of reduction. (See also Chapter 5, § 5.3.)

parts of the natural heritage, there are National Monuments to protect the cultural heritage. The United Nations Educational, Scientific and Cultural Organization's list of World Heritage Sites includes both those of cultural and those of natural importance. Once again, these parallels deserve more analysis than has been attempted here.

(iii) Can transformative values be used to formulate a complete aesthetic theory? This is an open question, though there is some reason for cautious optimism. Any aesthetic theory based on transformative value in this way will be at odds with views of art that rely entirely on shared commitments of members of a community that require no further justification.[4] Whether this constitutes a fatal objection to such a theory depends on whether such "institutional" views of art are found plausible. At the other extreme, appeal to transformative value is in accord with views that base aesthetics on some connection between aesthetic appreciation and emotion understood naturalistically.[5] In particular, if aesthetic response is mediated through the affective response, what the sensory input does is transform mental states, with conscious reflection playing no role at all.

Such an aesthetic theory based on transformative value works most easily with aesthetic universals. However, if it were capable only of accounting for aesthetic universals, it would not be of much use as a theory: contrary to Hume, aesthetic universals across cultures and historical eras seem to be relatively rare. Transformations of values brought on by some experience may well depend on an individual's past experiential history: this history dependence can account for variant aesthetic judgments that reflect individuals' cultural (and other) contexts.[6] Similarly, such an aesthetic theory is also consistent with views that regard art as social activity: such activity would be presumed to transform individual demand values in the relevant ways. Finally, from this perspective, similarities in the response to natural beauty (or the sublime) and to works of art are no puzzle: such similarities arise when a work of art produces transformations in the same way that natural beauty does. However, when a work of art transforms values, but does not do so in the same way as natural beauty or the sublime, then aesthetic appreciation occurs in a manner that is not similar

[4] See, for instance, the views of institutionalists such as George Dickie (e.g., Dickie 1974).

[5] For such an account of emotion, see Griffiths (1997).

[6] The situation with biodiversity is essentially the same: whether a particular experience of biodiversity transforms an individual may also depend on the past experience of that individual.

to the appreciation of nature. It is also possible, though it does not appear to be likely, that there can be a non-naturalistic account of aesthetic appreciation or taste based on transformative value. However, the motivation for attempting to develop such an account remains unclear.

(iv) What does the framework developed in this book show about the nature-culture divide that is often assumed in discussions of the problems associated with human interactions with nature? At no point were the discussions of this book dependent on making a distinction between culture and nature:[7] from a biological point of view, given the continuity in behavior from other animals to humans, the invocation of a culture-nature distinction, unless it is very carefully circumscribed, is a recipe for disaster. Moreover, making a sharp distinction and taking it to be valid is part of what spawns the myth of the golden age (see Chapter 2, § 2.2). It also spawns questions that are superficially difficult to answer and, moreover, irrelevant to the development of either a conservationist ethic or effective strategies for conservation: if humans are animals, human culture is analogous to animal cultures. Why should human cultural practices leading to extinctions be condemned any more than we condemn a predator species for consuming its prey? Why should human activities not be looked at as another facet of evolution, even if they lead to extinctions (especially since extinction, along with adaptation and speciation, is one of the three standard evolutionary processes)? These questions are not ultimately difficult to answer, but the answers are not illuminating.

Rather than address such questions, the approach developed here relies on two other explicit distinctions: (a) the distinction between anthropocentric and non-anthropocentric foundations for a conservationist ethic, and (b) the distinction between anthropogenic and non-anthropogenic mechanisms for extinctions (and other biotic changes). In both cases, the ultimate distinction is between the role played by human interests and actions, directly and indirectly, and all other factors. It does not matter whether the human interests and actions are considered natural or cultural; that question needs no answer in the present context. It has presumably been shown by example in this book that the development of an adequate framework for biodiversity conservation need not take any position on the alleged nature-culture divide (including whether the distinction even makes cognitive sense).

[7] For an opposing point of view about what an environmental ethic necessitates, see Soper (1995) and Woods (2001).

One question that was not explicitly mentioned in this list, but that underlies all of the listed questions, is whether an adequate account of transformative value can be given. Leaving aside the question of whether the concept of transformative value is coherent (as assumed in this book), the account given in Chapter 4 remained tentative. Two serious problems, the boundary and directionality problems, were explicitly noted there (§§ 4.5–4.6). It is less than clear that the solutions proposed there (§ 4.7) are fully adequate. However, short of a solution to these two problems, no defense of biodiversity conservation based on the transformative value of biodiversity can be successful. The issue here is not merely whether the specific account of transformative value developed in Chapter 4, based on transformations wrought on felt preferences or demand values, is fully adequate. For once, the details do not matter. The critical assumption is that biodiversity matters because of the way in which it transforms us as individuals: what exactly constitutes that transformation will distinguish different accounts of transformative value. Even at this level of generality, the force of the boundary and directionality problems remains no less compelling than before. Whether the normative ethical basis developed in this book is cogent will be determined by the future of accounts of transformative value.

8.2. THE SCIENCE OF BIODIVERSITY CONSERVATION

Turning to the foundations of conservation biology, there are at least the following six questions (and possibly others) that merit more philosophical attention. As before, the emphasis is on unresolved issues:

(i) what if the surrogacy problem has no solution? There have been so few systematic tests of the adequacy of surrogates that this is a very real possibility. There have been very few cases in which adequate surrogates have been found.[8] This may reflect only the fact that there have been relatively few attempts at explicit surrogacy analysis. However, even some of the successful cases are problematic, especially when the distributions of the true surrogates are not entirely based on survey results. For instance, in the case of Texas, the true surrogates used in the analysis were all of the vertebrate species. Their distributions were partially based on surveys and reports but also partially based on

[8] See, for example, the case of Texas (discussed in Chapter 6, § 6.3) and the results of Sarkar and colleagues (2005), which were obtained after this book was written.

models that used climatic and habitat characteristics. Thus there is a purely statistical reason why these distributions would be correlated with those of the environmental surrogates used. This is a methodological problem that must eventually be addressed if the theory of conservation biology is to be further developed. However, for the practical purpose of devising conservation strategies, since all that matters is that the true surrogates get adequately represented in a prioritized list of places, this is not a fatal flaw. True surrogates can be successfully selected using the estimator-surrogates, irrespective of the failure of statistical independence.

If, in general, adequate surrogates that can easily be assessed (such as environmental data and data from satellite images) cannot be found, then, at the very least, *rapid* biodiversity assessments cannot be performed. There seem to be only two ways out. (a) It may be possible to rectify the inadequacy of surrogates during conservation planning by setting higher targets for the representation of surrogates than would otherwise have been chosen. It is an open question how successfully this strategy can be used. (b) Some methodology other than the two discussed in Chapter 6 (§ 6.3) may be devised to select and test the adequacy of surrogates.[9] For instance – and this is admittedly far-fetched – there may be ways of bypassing the question of the representation of true surrogates by going directly from the estimator-surrogates to final conservation plans. Such a method would rather radically alter the framework for systematic conservation planning presented in Chapter 6 (§ 6.1).

(ii) What if the viability problem has no solution? At present, even singlepopulation viability models have so much structural uncertainty associated with them that their predictions are unreliable. But this should not be a reason for excessive pessimism. Population viability analysis is a little over two decades old; it is only now beginning to receive the kind of theoretical attention that many other ecological techniques have received over several generations. It is fairly safe to predict that many metapopulation and metacommunity (multispecies) viability models will be forthcoming over the next decade. Moreover, as noted in Chapter 6 (§ 6.4), population viability analysis is only one way to assess viability. Habitat-based viability models show considerable promise for situations in which population viability models are

[9] Sarkar and colleagues (2005) report some new attempts using complex tests of statistical associations.

either too complex to be tractable or requir
not be collected rapidly enough for conserv
individual-based, spatially explicit models a
new, predictively successful families of eco
in Chapter 6 (§ 6.4), these will include pre
of populations and communities – that is, in
conservation, of viability.

However, it may well be the case that the p
of some aspect of biodiversity fades into in
pogenic threats are taken into account. Ris
are designed to take these considerations int
simplest of circumstances, nontrivial and r
yet forthcoming.[10] Even given the future s
eling, the viability problem may remain re
conservation biology will have failed to re
logical science. In that case, the most ratio
conservation to take would be to proceed by
expert local knowledge. By and large, thi
practiced before the deliberate attempt to cr
tion biology. In a sense, the brave new scien
a medley of techniques (primarily for plac
relation to the North American developmen
of a new scientific discipline in the mid-19

(iii) What if the project of methodological inc
ological individualism, as noted in Chapter
ists' dream in ecology. There are many wit
sympathy for this dream. For some, this is u
tional issue: reductionism is supposed to re
toward the natural world. These positions
only emotional responses, merit no more at
work – than appeals to religious authority
Nevertheless, there are also good scientific
about the reductionist dream. In the biolo
traditionally been the discipline least ame
ods, simply because of the complexity of po
and the lack of simple equations describi
rately (such as Mendel's rules in evolution
that the systems to be modeled will prove

(iv)

[10] This is part of the general problem of the lack of predictabili

[11] This
prob

of their complexity, even in the face of the increased power of modern computers. It is also possible that the complexities involved are not merely theoretical or computational; it may be that some critical individual interactions are impossible to detect or estimate accurately in the field and thus impossible to model successfully (recall the discussion of Chapter 7, § 7.4). It may also be the case, though it does not appear to be likely, that there are metaphysical reasons, analogous to those in quantum mechanics and condensed matter physics, that individual-based explanation and prediction will ultimately fail in ecology.[11] Much of the optimism expressed in (ii) about the solution of the viability problem was based on the promise of individual-based models. Their failure may mean that the viability problem cannot be solved. However, once again, such pessimism is unwarranted at this point. Because of the complexities involved, individual-based models are most likely to fail at very large spatial scales. As long as they are successful at smaller spatial scales, however, they may be sufficient to assess viabilities to the extent required for biodiversity conservation planning.

What if multiple criteria can almost never be synchronized? There are two aspects to this problem, one that is foundational and the other empirical. A significant foundational problem arises when the MCS procedure generates unwieldy sets of non-dominated solutions. There are several ways in which these sets may be unwieldy, but the most important of them is when sets of non-dominated solutions are intractably large for practical planning purposes. In such a situation, further refinement of sets, by choosing only some of the non-dominated alternatives, is ultimately arbitrary because either (a) refinement has to be carried out by making counterfactual assumptions, for instance, that the alternatives and criteria can be metrically ordered, or (b) special assumptions are now introduced to exclude some alternatives. The best that can be hoped for is that these special factors reflect intangible values that can be reasoned about, even if they cannot be incorporated into an algorithmic procedure.

Equally important from the practical point of view, though not of the same foundational importance, is the case when there is sufficient disharmony between the valuations along the different criteria that different non-dominated solutions appear equally unattractive to different

[11] ... is less likely simply because, so far, the reductionist program has encountered far fewer ... problems in biology than in physics – see Sarkar (1998a).

individuals involved in conservation planning. Empirically, what is acceptable to loggers will usually be anathema to wilderness advocates; what has low social cost may have very high economic cost. All of these criteria may lead to diametrically opposed alternatives. There may be no compromise resolution of these dilemmas. This may be a disastrous situation for biodiversity conservation. However, it seems unreasonable to demand that conservation biology provide an answer in such cases. These disagreements reflect conflicts of values that are features of the world. Their resolution will require deliberations about the relative importance of the norms they incorporate, beyond any issue that scientific analyses can or should aim to resolve.

(v) What if a satisfactory systematic method to incorporate uncertainties can never be worked out? Much of what was said in Chapter 7 (§ 7.6) about coping with uncertainty constituted little more than promissory notes. For both types of uncertainty, those associated with data and those that arise for of structural reasons, the methods available at present do not provide significant guidance for decision making. Moreover, determining the uncertainties associated with any final outcome of a conservation planning process may well prove to be an intractable problem because of the uncertainties involved at every stage of the process: (a) uncertainties associated with data gathering and analysis (for place prioritization and surrogacy assessment), (b) structural uncertainties associated with the ecological models involved (for viability analysis), and (c) a lack of control over the uncertainties of the sociopolitical considerations that are necessary to accommodate incommensurable values. New methods of uncertainty analysis will almost certainly have to be devised. As suggested in Chapter 7 (§ 7.6), Bayesian statistics may provide a framework for combining all of these uncertainties into a single result. However, this remains to be systematically proved in practice. Perhaps this is where philosophy of science has the most to contribute to biodiversity conservation.

(vi) Is biodiversity even the appropriate target of concern for nature? Angermeier and Karr have forcefully argued that biological integrity rather than biodiversity should be the target of concern; the former is supposed to "refer to a system's wholeness, including presence of all appropriate elements and occurrences of all processes at appropriate rates."[12] Not only is integrity supposed to be more inclusive than diversity, because it includes processes as targets of concern, it is also

[12] Angermeier and Karr (1994), p. 692. See also Karr (1991).

supposed to provide a better safeguard against anthropogenic change: human interference can increase the biodiversity of a region but not, according to Angermeier and Karr, its biological integrity.[13] The definition continues: "Whereas diversity is a collective property of system elements, integrity is a synthetic property of the system. Unlike diversity, which can be expressed simply as the number of kinds of items, integrity refers to conditions under little or no influence from human actions; a biota [*sic*] with high integrity reflects natural evolutionary and biogeographic processes."[14] Once again, we seem to be operating under the aegis of the myth of the golden age: humans are not part of the biological world; desired landscapes must be those that exclude the effects of human activity, whether or not such landscapes exist. Claims such as this ignore the fact that whether human activities harm either biodiversity or biological integrity is an empirical question (recall the discussions of Chapter 2, § 2.2). Unless it is taken as a matter of faith that human presence in an ecosystem is itself undesirable, the role of humans in an ecosystem must be assessed empirically before any claim can be made about the desirability of human presence.

However, there is also the other and more important aspect of this appeal to integrity: the move to include processes along with entities. This has the benefit that features such as endangered biological phenomena, noted as a problem in the discussion of Chapter 6 (§ 6.5), naturally enter into considerations of what should be conserved. Nevertheless, in general, the appeal to biological integrity has so far suffered from the fatal problem that the concept appears to be impossible to operationalize for use in the field. Even if it were accepted that biological integrity is a desirable target – along with, if not instead of, biodiversity – what is the "appropriate" level for each process? Vague appeals to evolutionary and biogeographic processes are of no help: these are neither constant nor necessarily equilibrating processes.[15]

Both evolution and biogeography are about change, sometimes radical change that generates the variety of life around us. Moreover, viability analyses for biodiversity must take into account those aspects of processes that are critical for the survival of the biotic features

[13] The rationale for this claim remains unclear: the Keoladeo case (Chapter 2, 2.3) is one in which human interactions maintain what, *intuitively*, should be regarded as the integrity of the system as bird habitat. But this objection is overruled by Angermeier and Karr in their *definition* of biological integrity – see the text.

[14] Angermeier and Karr (1994), p. 692.

[15] It seems to assume the mythical balance of nature – see Chapter 5 (§ 5.1).

> that are of interest. Nevertheless, there is perhaps one valuable lesson to be learned from the appeal to biological integrity, for instance, in the context of the conservation of endangered biological phenomena: biodiversity alone, especially after it is operationalized in the way indicated in Chapter 6 (§ 6.5), does not capture all that is valuable and worthy of conservation in the natural world. Our concern for the natural world should not stop with biodiversity; rather, it should only begin there, as we try to understand what it is about nature that we value or should value. (That should be a goal of environmental philosophy beyond concern for biodiversity conservation.)

It is quite possible that the conservation biology of the future will be as different from that of today as the framework discussed in Chapter 6 is different from that first formulated in the North American context in the 1980s. It is almost certain that the techniques discussed in this book will be superseded by others: that has been the fate of early developments in every other science, and there is no reason to expect – or desire – that conservation biology will prove to be an exception. Of the four problems discussed in Chapter 6, only the place prioritization problem has been addressed successfully enough for there to be any confidence that some of the basic approaches will be long-lasting. There can as yet be no similar confidence about our attempted solutions of the surrogacy, viability, and feasibility problems. At a more general level, it will be interesting to see if the current schematization of these problems survives or whether the development of new techniques will lead to a radically different conceptual structure for conservation biology. In particular, the sociopolitical considerations that have to be addressed in attempts to solve the feasibility problem may well result in radical structural modifications.

Perhaps this is the place to emphasize, one last time, that biodiversity conservation is as much a sociopolitical issue as a scientific one. Even situations as simple as the conservation of Olive Ridley turtles (*Lepidochelys olivacea*) in the face of poverty and inadequate availability of protein in Orissa (India) (see Chapter 1) cannot easily be incorporated into the systematic framework introduced in Chapter 6 but nevertheless must be negotiated for successful conservation. Conservation biologists can forget the sociopolitical context of biodiversity conservation only at the risk of losing what remains of our biological heritage. Our final open question is how much of the social scientific and other sociopolitical analyses must be incorporated into conservation biology for there to emerge a truly reliable science dedicated to the conservation of the biodiversity around us.

References

Ackery, P. R., and Vane-Wright, R. I. 1984. *Milkweed Butterflies*. Ithaca, NY: Cornell University Press.

Aerts, R., and Chapin, F. S., III. 2000. "The Mineral Nutrition of Wild Plants Revisited: A Re-Evaluation of Processes and Patterns." *Advances in Ecological Research* **30**: 1–67.

Agarwal, A. 1992. "Sociological and Political Constraints to Biodiversity Conservation." In Sandlund, O. T., Hinder, K., and Brown, A. H. D., eds., *Conservation of Biodiversity for Sustainable Development*. Oslo: Scandinavian University Press, pp. 293–302.

Aggarwal, A., Garson, J., Margules, C. R., Nicholls, A. O., and Sarkar, S. 2000. *ResNet Ver 1.1 Manual*. Austin: University of Texas Biodiversity and Biocultural Conservation Laboratory.

Aiken, W. 1992. "Human Rights in an Ecological Era." *Environmental Values* **1**: 191–203.

Alihan, M. A. 1938. *Social Ecology: A Critical Analysis*. New York: Columbia University Press.

Alvarez, L. W., Alvarez, W., Asaro, F., and Michel, H. V. 1980. "Extraterrestrial Cause for the Cretaceous-Tertiary Extinction." *Science* **208**: 1095–1098.

Amato, G., Rabinowitz, A., and Egan, M. G. 1999. "A New Species of Muntjac, *Muntiacus putaoensis* (Artiodactyla: Cervidae from Northern Myanmar)." *Animal Conservation* **2**: 1–7.

Andelman, S. J., and Fagan, W. F. 2000. "Umbrellas and Flagships: Efficient Conservation Surrogates, or Expensive Mistakes?" *Proceedings of the National Academy of Sciences (USA)* **97**: 5954–5959.

Angermeier, P. L., and Karr, J. R. 1994. "Biological Integrity versus Biological Diversity as Policy Directives." *BioScience* **44**: 690–697.

Araújo, M. B., Densham, P. J., Lampinen, R., Hagemeijer, W. J. M., Mitchell-Jones, A. J., and Gase, J. P. 2001. "Would Environmental Diversity Be a Good Surrogate for Species Diversity?" *Ecography* **24**: 103–110.

Arrhenius, O. 1921. "Species and Area." *Journal of Ecology* **9**: 95–99.

Arrow, K. J., and Raynaud, H. 1986. *Social Choice and Multicriterion Decision-Making*. Cambridge, MA: MIT Press.

Arthur, J. L., Hachey, M., Sahr, K., Huso, M., and Kiester, A. R. 1997. "Finding All Optimal Solutions to the Reserve Site Selection Problem: Formulation and Computational Analysis." *Environmental and Ecological Statistics* **4**: 153–165.

Asteris, G., and Sarkar, S. 1996. "Bayesian Procedures for the Estimation of Mutation Rates from Fluctuation Experiments." *Genetics* **142**: 313–326.

Atfield, R. 1987. *A Theory of Value and Obligation*. London: Croom Helm.

Athanasiou, T., and Baer, P. 2003. *Dead Heat: Global Justice and Global Warming*. New York: Seven Sisters Press.

Austin, M. P., and Heyligers, P. C. 1989. "Vegetation Survey Design for Conservation: Gradsect Sampling of Forests in North-eastern New South Wales." *Biological Conservation* **50**: 13–32.

Austin, M. P., and Margules, C. R. 1986. "Assessing Representativeness." In Usher, M. B., ed., *Wildlife Conservation Evaluation*. London: Chapman and Hall, pp. 45–67.

Bahadur, K. N. 1986. "Bamboos." In Hawkins, R. E., ed., *Encyclopedia of Indian Natural History*. New Delhi: Oxford University Press, pp. 30–38.

Balmford, A., Bruner, A., Cooper, P., Constanza, R., Farber, S., Green, R. E., Jenkins, M., Jefferiss, P., Jessamy, V., Madden, J., Munro, K., Myers, N., Naeem, S., Paavola, J., Rayment, M., Rosendo, S., Roughgarden, J., Trumper, K., and Turner, R. K. 2002. "Economic Reasons for Conserving Nature." *Nature* **297**: 950–953.

Barbier, E. B., Brown, G., Dalmazzone, S., Folke, C., Gadgil, M., Hanley, N., Holling, C. S., Lesser, W. H., Mäler, K.-G., Mason, P., Panayotou, T., Perrings, C., Turner, R. K., and Wells, M. 1995. "The Economic Value of Biodiversity." In Heywood, V. H., ed., *Global Biodiversity Assessment*. Cambridge, UK: Cambridge University Press, pp. 824–914.

Barrow, C. J. 2000. *Social Impact Assessment: An Introduction*. New York: Oxford University Press.

Batchelor, M., and Brown, K., eds. 1994. *Buddhism and Ecology*. New Delhi: Motilal Banarsidass.

Bateman, I. J., and Willis, K. G., eds. 1999. *Valuing Environmental Preferences*. Oxford: Oxford University Press.

Beissinger, S. R. 2002. "Population Viability Analysis: Past, Present, Future." In Beissinger, S. R., and McCullough, D. R., eds., *Population Viability Analysis*. Chicago: University of Chicago Press, pp. 5–17.

Beissinger, S. R., and McCullough, D. R., eds. 2002. *Population Viability Analysis*. Chicago: University of Chicago Press.

Beissinger, S. R., and Westphal, M. I. 1998. "On the Use of Demographic Models of Population Viability in Endangered Species Management." *Journal of Wildlife Management* **62**: 821–841.

Benyus, J. M. 1997. *Biomimicry*. New York: William Morrow.

Bevis, W. W. 1995. *Borneo Log: The Struggle for Sarawak's Forests*. Seattle: University of Washington Press.

Biehl, J., and Staudenmaier, P. 1995. *Ecofascism: Lessons from the German Experience*. Edinburgh: AK Press.

Bjørklund, I. 1990. "Sámi Reindeer Pastoralism as an Indigenous Resource Management System in Northern Norway: A Contribution to the Common Property Debate." *Development and Change* **21**: 75–86.

Bogetoft, P., and Pruzan, P. 1997. *Planning with Multiple Criteria: Investigation, Communication and Choice.* Copenhagen: Copenhagen Business School Press.

Bolker, J. A. 1995. "Model Systems in Developmental Biology." *BioEssays* **17**: 451–455.

Bookchin, M. 1995. *The Philosophy of Social Ecology: Essays on Dialectical Naturalism.* Montreal: Black Rose Books.

Borgmann, A. 1995. "The Nature of Reality and the Reality of Nature." In Soulé, M. E., and Lease, G., eds. *Reinventing Nature? Responses to Postmodern Deconstruction.* Washington, DC: Island Press, pp. 31–45.

Bormann, F. H., and Likens, G. E. 1979a. "Catastrophic Disturbance and the Steady State in Northern Hardwood Forests." *American Scientist* **67**: 660–669.

Bormann, F. H., and Likens, G. E. 1979b. *Pattern and Process in a Forested Ecosystem.* Berlin: Springer-Verlag.

Boulding, K. 1966. "The Economics of the Coming Spaceship Earth." In Jarrett, H., ed., *Environmental Quality in a Growing Economy.* Baltimore: Johns Hopkins University Press, pp. 3–14.

Boyce, M. S., Meyer, J. S., and Irwin, L. 1994. "Habitat-Based PVA for the Northern Spotted Owl." In Fletcher, D. J., and Manly, B. F. J., eds., *Statistics in Ecology and Environmental Monitoring.* Dunedin, New Zealand: University of Otago Press, pp. 63– 85.

Bramwell, A. 1989. *Ecology in the 20th Century: A History.* New Haven, CT: Yale University Press.

Brower, L. P., and Malcolm, S. B. 1991. "Animal Migrations: Endangered Phenomena." *American Zoologist* **31**: 265–276.

Brown, J. H., and Lomolino, M. V. 1989. "Independent Discovery of the Equilibrium Theory of Island Biogeography." *Ecology* **70**: 1954–1957.

Buss, L. 1987. *The Evolution of Individuality.* Princeton, NJ: Princeton University Press.

Cabeza, M., and Moilenan, A. 2001. "Design of Reserve Networks and the Persistence of Biodiversity." *Trends in Ecology and Evolution* **16**: 242–248.

Callicott, J. B. 1980. "Animal Liberation: A Triangular Affair." *Enviromental Ethics* **2**: 331–338.

Callicott, J. B. 1986. "On the Intrinsic Value of Nonhuman Species." In Norton, B. G., ed., *The Preservation of Species: The Value of Biological Diversity.* Princeton, NJ: Princeton University Press, pp. 138–172.

Callicott, J. B. 1989. *In Defense of the Land Ethic: Essays in Environmental Philosophy.* Albany: State University of New York Press.

Callicott, J. B. 1991a. "The Good Old-Time Wilderness Religion." *Environmental Professional* **13**: 378–379.

Callicott, J. B. 1991b. "The Wilderness Idea Revisited: The Sustainable Development Alternative." *Environmental Professional* **13**: 235–247.

Carr, A. 1984. *The Sea Turtle: So Excellent a Fishe.* Austin: University of Texas Press.

Carson, R. 1962. *Silent Spring.* Boston: Houghton-Mifflin.

Caufield, C. 1984. *In the Rainforest: Report from a Strange, Beautiful, Imperiled World.* Chicago: University of Chicago Press.

Caughley, G. 1994. "Directions in Conservation Biology." *Journal of Animal Ecology* **63**: 215–244.

Caughley, G., and Gunn, A. 1996. *Conservation Biology in Theory and Practice.* Boston: Blackwell Science.

Church, R. L., Stoms, D. M., and Davis, F. W. 1996. "Reserve Selection as a Maximal Covering Location Problem." *Biological Conservation* **76**: 105–112.

Cockburn, A., and Ridgeway, J., eds. 1979. *Political Ecology.* New York: Times Books.

Cocks, K. D., and Baird, I. A. 1989. "Using Mathematical Programming to Address the Multiple Reserve Selection Problem: An Example from the Eyre Peninsula, South Australia." *Biological Conservation* **49**: 113–130.

Cohen, G. A. 2000. *If You're an Egalitarian, How Come You're So Rich?* Cambridge, MA: Harvard University Press.

Cohen, J. E. 1995. *How Many People Can the Earth Support?* New York: Norton.

Connor, E. F., and McCoy, E. D. 1979. "The Statistics and Biology of the Species-Area Relationship." *American Naturalist* **113**: 791–833.

Cox, S. J. B. 1985. "No Tragedy on the Commons." *Environmental Ethics* **7**: 49–61.

Cronon, W. 1996a. "Introduction: In Search of Nature." In Cronon, W., ed., *Uncommon Ground: Rethinking the Human Place in Nature.* New York: Norton, pp. 23–56.

Cronon, W. 1996b. "The Trouble with Wilderness; or, Getting Back to the Wrong Nature." In Cronon, W., ed., *Uncommon Ground: Rethinking the Human Place in Nature.* New York: Norton, pp. 69–90.

Cronon, W., ed. 1996c. *Uncommon Ground: Rethinking the Human Place in Nature.* New York: Norton.

Denevan, W. M. 1992. "The Pristine Myth: The Landscape of the Americas in 1492." *Annals of the Association of American Geographers* **82**: 369–385.

Dennis, B. 1996. "Discussion: Should Ecologists Become Bayesians?" *Ecological Applications* **6**: 1095–1103.

Diamond, J. M. 1975a. "Assembly of Species Communities." In Cody, M. L., and Diamond, J. M., eds., *Ecology and the Evolution of Communities.* Cambridge, MA: Harvard University Press, pp. 342–459.

Diamond, J. M. 1975b. "The Island Dilemma: Lessons of Modern Biogeographic Studies for the Design of Natural Reserves." *Biological Conservation* **7**: 129–146.

Diamond, J. M. 1976. "Island Biogeography and Conservation: Strategy and Limitations." *Science* **193**: 1027–1029.

Diamond, J. M., and May, R. M. 1976. "Island Biogeography and the Design of Natural Reserves." In May, R. M., ed., *Theoretical Ecology: Principles and Applications.* Oxford: Blackwell, pp. 163–186.

Dickie, G. 1974. *Art and the Aesthetic: An Institutional Analysis.* Ithaca, NY: Cornell, University Press.

Dyer, J. S., and Sarin, R. K. 1979. "Measurable Multiattribute Value Functions." *Operations Research* **27**: 810–822.

Eagleton, T. 1991. *Ideology: An Introduction.* London: Verso.

Eagleton, T., ed. 1994. *Ideology.* London: Longmans.

Easterbrook, G. 1995. *A Moment on Earth: The Coming Age of Environmental Optimism.* New York: Viking.

Ebenhard, T. 1988. "Introduced Birds and Mammals and Their Ecological Effects." *Swedish Wildlife Research* **13**: 1–107.

Ehrenfeld, D. W. 1976. "The Conservation of Non-Resources." *American Scientist* **64**: 648–656.

Ehrlich, P. R., and Ehrlich, A. 1968. *The Population Bomb*. New York: Ballantine.

Ehrlich, P. R., and Ehrlich, A. 1981. *Extinction: The Causes and Consequences of the Disappearance of Species*. New York: Random House.

Elliott, R. 2001. "Normative Ethics." In Jamieson, D., ed., *A Companion to Environmental Philosophy*. Malden, UK: Blackwell, pp. 175–191.

Ellison, A. M. 1996. "An Introduction to Bayesian Inference for Ecological Research and Environmental Decision-Making." *Ecological Applications* **6**: 1036–1046.

Ereshefsky, M., ed. 1992. *The Units of Evolution: Essays on the Nature of Species*. Cambridge, MA: MIT Press.

Faber, D. 1993. *Environment under Fire: Imperialism and the Ecological Crisis in Central America*. New York: Monthly Review Press.

Fahrig, L. 2003. "Effects of Habitat Fragmentation on Biodiversity." *Annual Review of Ecology, Evolution and Systematics* **34**: 487–515.

Faith, D. P. 1995. *Biodiversity and Regional Sustainability Analysis*. Lyneham: CSIRO Division of Wildlife and Ecology.

Faith, D. P., and Walker, P. A. 1996. "How Do Indicator Groups Provide Information about the Relative Biodiversity of Different Sets of Areas?: On Hotspots, Complementarity, and Pattern-based Approaches." *Biodiversity Letters* **3**: 18–25.

Fanon, F. 1963. *The Wretched of the Earth*. Harmondsworth: Penguin.

Fayrer-Hosken, R. A., Grobler, D., Van Altena, J. J., Bertschinger, H. J., and Kirkpatrick, J. F. 2000. "Immunocontraception of African Elephants." *Nature* **407**: 149.

Fayrer-Hosken, R. A., Grobler, D., Van Altena, J. J., Bertschinger, H. J., and Kirkpatrick, J. F. 2001. "African Elephants and Contraception." *Nature* **411**: 766.

Feinberg, J. 1974. "The Rights of Animals and Unborn Generations." In Blackstone, W. I., ed., *Philosophy and the Enviornmental Crisis*. Athens, University of Georgia Press, pp. 43–68.

Ferrier, S., and Watson, G. 1997. *An Evaluation of the Effectiveness of Environmental Surrogates and Modelling Techniques in Predicting the Distribution of Biological Diversity: Consultancy Report to the Biodiversity Convention and Strategy Section of the Biodiversity Group, Environment Australia*. Arimidale: Environment Australia.

Ferry, L. 1995. *The New Ecological Order*. Chicago: University of Chicago Press.

Fisher, J. A. 2001. "Aesthetics." In Jamieson, D., ed., *A Companion to Environmental Philosophy*. Malden, UK: Blackwell, pp. 264–276.

Foley, P. 1994. "Predicting Extinction Times from Environmental Stochasticity and Carrying Capacity." *Conservation Biology* **8**: 124–137.

Foley, P. 1997. "Extinction Models for Local Populations." In Hanski, I. A., and Gilpin, M. E., eds., *Metapopulation Biology*. New York: Academic Press, pp. 215–246.

Foreman, D. 1991. "Second Thoughts of an Eco-Warrior." In Chase, S., ed., *Defending the Earth: A Dialogue between Murray Bookchin and Dave Foreman*. Boston: South End Press, pp. 107–119.

Fowler, L. E. 1979. "Hatching Success and Nest Predation in the Green Sea Turtle, *Chelonia mydas*, at Tortuguero, Costa Rica." *Ecology* **60**: 946–955.

Fox, W. 1993. "What Does the Recognition of Intrinsic Value Entail?" *Trumpeter* **10**: 101.

Frankena, W. K. 1939. "The Naturalistic Fallacy." *Mind* **48**: 464–477.

Fryxell, J. M., and Lundberg, P. 1998. *Individual Behavior and Community Dynamics*. London: Chapman and Hall.

Funch, P., and Kristensen, R. M. 1995. "*Cycliophora* Is a New Phylum with Affinities to *Entoprocta* and *Ectoiprocta*." *Nature* **378**: 711–714.

Gadgil, M., and Berkes, F. 1991. "Traditional Resource Management Systems." *Resource Management and Optimization* **8**: 127–141.

Gadgil, M., and Guha, R. 1992. *This Fissured Land: An Ecological History of India*. New Delhi: Oxford University Press.

Gadgil, M., and Guha, R. 1995. *Ecology and Equity: The Use and Abuse of Nature in Contemporary India*. New Delhi: Penguin.

Garey, M. R., and Johnson, D. S. 1979. *Computers and Intractability: A Guide to the Theory of NP-Completeness*. San Francisco: W. H. Freeman.

Garson, J., Aggarwal, A., and Sarkar, S. 2002. "Birds as Surrogates for Biodiversity: An Analysis of a Data Set from Southern Québec." *Journal of Biosciences* **27** (S2): 347–360.

Gaston, K. J., and May, R. M. 1992. "The Taxonomy of Taxonomists." *Nature* **356**: 281–282.

Gause, G. F. 1934. *The Struggle for Existence*. Baltimore: Williams and Wilkins.

George, W. 1981. "Wallace and His Line." In Whitmore, T. C., ed., *Wallace's Line and Plate Tectonics*. Oxford: Clarendon Press, pp. 3–8.

Giao, P. M., Tuoc, D., Dung, V. V., Wikramanayake, E., Amato, G., Arctander, P., and MacKinnon, J. R. 1998. "Description of *Muntiacus truongsonensis*, a New Species of Muntjac (Artiodactyla: Muntiacidae) from Central Vietnam, and Implications for Conservation." *Animal Conservation* **1**: 61–68.

Gilbert, F. S. 1980. "The Equilibrium Theory of Island Biogeography: Fact or Fiction?" *Journal of Biogeography* **7**: 209–235.

Gilpin, M. E., and Soulé, M. E. 1986. "Minimum Viable Populations: Processes of Species Extinction." In Soulé, M. E., ed., *Conservation Biology: The Science of Scarcity and Diversity*. Sunderland, MA: Sinauer, pp. 19–34.

Gleason, H. A. 1922. "On the Relation between Species and Area." *Ecology* **3**: 158–162.

Golley, F. B. 1993. *A History of the Ecosystem Concept in Ecology: More than the Sum of the Parts*. New Haven, CT: Yale University Press.

Goodman, D. 1975. "The Theory of Diversity-Stability Relationships in Ecology." *Quarterly Review of Biology* **50**: 237–266.

Goodman, D. 2002. "Predictive Bayesian Population Viability Analysis: A Logic for Listing Criteria, Delisting Criteria, and Recovery Plans." In Beissinger, S. R., and McCullough, D. R., eds., *Population Viability Analysis*. Chicago: University of Chicago Press, pp. 447–469.

Goodpaster, K. 1978. "On Being Morally Considerable." *Journal of Philosophy* **75**: 308–325.

Griffiths, P. E. 1997. *What Emotions Really Are: The Problem of Psychological Categories*. Chicago: University of Chicago Press.

Grimm, V., and Wissel, C. 1997. "Babel, or the Ecological Stability Discussions: An Inventory and Analysis of Terminology and a Guide for Avoiding Confusions." *Oecologia* **109**: 323–334.

Groombridge, B., ed. 1992. *Global Biodiversity: Status of the Earth's Living Resources*. London: Chapman and Hall.

Groves, C. R., Jensen, D. B., Valutis, L. L., Redford, K. H., Shaffer, M. L., Scott, J. M., Baumgartner, J. V., Higgins, J. V., Beck, M. W., and Anderson, M. G. 2002.

"Planning for Biodiversity Conservation: Putting Conservation Science into Practice." *BioScience* **52**: 499–512.

Guha, R. 1989. "Radical American Environmentalism and Wilderness Preservation: A Third World Critique." *Environmental Ethics* **11**: 71–83.

Guha, R., ed. 1994. *Social Ecology*. Delhi: Oxford University Press.

Haila, Y. 1986. "On the Semiotic Dimension of Ecological Theory: The Case of Island Biogeography." *Biology and Philosophy* **1**: 377–387.

Haila, Y., 1997. "Trivialization of Critique in Ecology." *Biology and Philosophy* **12**: 109–118.

Haila, Y., and Levins, R. 1992. *Humanity and Nature: Ecology, Science and Society*. London: Pluto Press.

Haila, Y., and Margules, C. R. 1996. "Survey Research in Conservation Biology." *Ecography* **19**: 323–331.

Hanski, I. 1999. *Metapopulation Ecology*. Oxford: Oxford University Press.

Hardin, G. 1968. "The Tragedy of the Commons." *Science* **162**: 1243–1248.

Hardin, G. 1995. *The Immigration Dilemma: Avoiding the Tragedy of the Commons*. Washington: Federation for American Immigration Reform.

Harrison, S. 1991. "Population Growth, Land Use and Deforestation in Costa Rica." *Interciencia* **16**: 83–93.

Hecht, S., and Cockburn, A. 1990. *The Fate of the Forest: Developers, Destroyers and Defenders of the Amazon*. New York: Harper Perennial.

Hedrick, P. W., Lacy, R. C., Allendorf, F. W., and Soulé, M. E. 1996. "Directions in Conservation Biology: Comments on Caughley." *Conservation Biology* **10**: 1312–1320.

Heinzerling, L., and Ackerman, F. 2002. *Pricing the Priceless: Cost-Benefit Analysis of Environmental Protection*. Washington, DC: Georgetown University Law Center.

Higgs, A. J. 1981. "Island Biogeography Theory and Nature Reserve Design." *Journal of Biogeography* **8**: 117–124.

Hobbs, R. J., and Huenneke, L. F. 1992. "Disturbance, Diversity, and Invasion – Implications for Conservation." *Conservation Biology* **6**: 324–337.

Holling, C. S. 1973. "Resilience and Stability of Ecological Systems." *Annual Review of Ecology and Systematics* **4**: 1–23.

Holling, C. S., ed. 1978. *Adaptive Environmental Assessment and Management*. Chichester: Wiley.

Horgan, J. 1995. "The New Social Darwinists." *Scientific American* **273**(4): 174–181.

Horgan, J. 1999. *The Undiscovered Mind: How the Human Brain Defies Replication, Medication, and Explanation*. New York: Free Press.

Horwich, C., and Urbach, P. 1993. *Scientific Reasoning: The Bayesian Approach*. La Salle, IL: Open Court.

Howard, P. C., Viskanic, P., Davenport, T. R. B., Kigenyi, F. W., Baltzer, M., Dickinson, C. J., Lwanga, J. S., Matthews, R. A., and Balmford, A. 1998. "Complementarity and the Use of Indicator Groups for Reserve Selection in Uganda." *Nature* **394**: 472–475.

Hume, D. 1948 [1779]. *Dialogues Concerning Natural Religion*. New York: Hafner.

Hume, D. 1972 [1737]. *A Treatise on Human Nature*. London: Fontana.

Huston, M., DeAngelis, D., and Post, W. 1988. "New Computer Models Unify Ecological Theory." *BioScience* **38**(10): 682–691.

International Union for the Conservation of Nature (IUCN). 1980. *World Conservation Strategy: Living Resource Conservation for Sustainable Development*. Gland, Switzerland: International Union for the Conservation of Nature.

International Union for the Conservation of Nature (IUCN). 1983. *Parks and Life: Report of the IVth World Congress on National Parks and Protected Areas*. Gland, Switzerland: International Union for the Conservation of Nature.

Jablonski, D. 1993. "Mass Extinctions: New Answers, New Questions." In Kaufman, L., and Mallory, K., eds., *The Last Extinction*, 2nd ed. Cambridge, MA: MIT Press, pp. 47–68.

Jablonski, D. 1995. "Extinctions in the Fossil Record." In Lawton, J. H., and May, R. M., eds. *Extinction Rates*. Oxford: Oxford University Press, pp. 25–44.

Jaffe, M. 1994. *And No Birds Sing: The Story of an Ecological Disaster in a Tropical Paradise*. New York: Simon and Schuster.

Janzen, D. H. 1986. "The Future of Tropical Ecology." *Annual Review of Ecology and Systematics* **17**: 305–324.

Jones, B. D. 1999. "Bounded Rationality." *Annual Review of Political Science* **2**(1): 297–321.

Justus, J., and Sarkar, S. 2002. "The Principle of Complementarity in the Design of Reserve Networks to Conserve Biodiversity: A Preliminary History." *Journal of Biosciences* **27** (S2): 421–435.

Karr, J. R. 1991. "Biological Integrity: A Long Neglected Aspect of Water Resource Management." *Ecological Applications* **1**: 66–84.

Kasischke, E. S., Christensen, N. L., and Stocks, B. J. 1995. "Fire, Global Warming, and the Carbon Balance of Boreal Forests." *Ecological Applications* **5**: 437–451.

Kass, L. R., and Wilson, J. O. 1998. *The Ethics of Human Cloning*. Washington, DC: AEI Press.

Kass, R. E., and Raftery, A. E. 1995. "Bayes Factors." *Journal of the American Statistical Association* **90**: 773–795.

Keeney, R. L., and Raiffa, H. 1993. *Decisions with Multiple Alternatives: Preferences and Value Tradeoffs*. Cambridge: Cambridge University Press.

Keller, D. R., and Golley, F. B., eds. 2000. *The Philosophy of Ecology: From Science to Synthesis*. Athens: University of Georgia Press.

Kellert, S. R. 1996. *The Value of Life: Biological Diversity and Human Society*. Washington, DC: Island Press.

Kellog, E. A., and Shaffer, H. B. 1993. "Model Organisms in Evolutionary Studies." *Systematic Biology* **42**: 409–414.

Kingsland, S. E. 1985. *Modeling Nature: Episodes in the History of Population Ecology*. Chicago: University of Chicago Press.

Kingsland, S. 2002. "Designing Nature Reserves: Adapting Ecology to Real-World Problems." *Endeavour* **26**: 9–14.

Kirkpatrick, J. B. 1983. "An Iterative Method for Establishing Priorities for the Selection of Nature Reserves: An Example from Tasmania." *Biological Conservation* **25**: 127–134.

Kitcher, P. 1985. *Vaulting Ambition: Sociobiology and the Quest for Human Nature*. Cambridge, MA: MIT Press.

Korsgaard, C. 1983. "Two Distinctions in Goodness." *Philosophical Review* **92**: 169–195.

Krieger, M. 1973. "What's Wrong with Plastic Trees?" *Science* **179**: 446–453.

Kuhn, T. 1962. *The Structure of Scientific Revolutions*. Chicago: University of Chicago Press.

Lacy, R. C., and Miller, P. S. 2002. "Incorporating Human Populations and Activities into Population Viability Analysis." In Beissinger, S. R., and McCullough, D. R., eds., *Population Viability Analysis*. Chicago: University of Chicago Press, pp. 490–510.

Lear, L. 1997. *Rachel Carson: Witness for Nature*. New York: Henry Holt.

Lee, M. F. 1995. *Earth First! Environmental Apocalypse*. Syracuse, NY: Syracuse University Press.

Lehman, C. L., and Tilman, D. 2000. "Biodiversity, Stability, and Productivity in Competitive Communities." *American Naturalist* **156**: 534–552.

Leopold, A. 1949. *A Sand County Almanac*. London: Oxford University Press.

Leopold, A. 1950. "Conservation." In Leopold, L., ed., *Round River: From the Journals of Aldo Leopold*. New York: Oxford University Press, pp. 145–157.

Levins, R. 1966. "The Strategy of Model Building in Population Biology." *American Scientist* **54**: 421–431.

Levins, R. 1974. "The Qualitative Analysis of Partially Specified Systems." *Annals of the New York Academy of Sciences* **231**: 123–138.

Levins, R. 1975. "Evolution in Communities Near Equilibrium." In Cody, M. L., and Diamond, J. M., eds. *Ecology and the Evolution of Communities*. Cambridge, MA: Harvard University Press, pp. 16–50.

Levins, R. 1993. "A Response to Orzack and Sober: Formal Analysis and the Fluidity of Science." *Quarterly Review of Biology* **68**: 547–555.

Lewis, M. 2003. *Inventing Global Ecology: Tracking the Biodiversity Ideal in India, 1945–1997*. New Delhi: Orient Longman.

Lewis, W. H., and Elvin-Lewis, M. P. F. 1977. *Medical Botany*. New York: John Wiley.

Lewontin, R. C. 1969. "The Meaning of Stability." *Brookhaven Symposia in Biology* **22**: 13–24.

Liem, D. S. 1973. "A New Genus of Frog of the Family Leptodactylidae from SE Queensland, Australia." *Memoirs of the Queensland Museum* **16**: 459–470.

Light, A., ed. 1998. *Social Ecology after Bookchin*. New York: Guilford Press.

Lomborg, B. 2001. *The Skeptical Environmentalist*. Cambridge: Cambridge University Press.

Loomis, J. B., and White, D. S. 1996. "Economic Benefits of Rare and Endangered Species: Summary and Meta-analysis." *Ecological Economics* **18**: 197–206.

Lovejoy, T. E. 1986. "Species Leave the Ark One by One." In Norton, B. G., ed., *The Preservation of Species*. Princeton, NJ: Princeton University Press, pp. 13–27.

MacArthur, R. H. 1965. "Patterns of Species Diversity." *Biological Review* **40**: 510–533.

MacArthur, R. H., and Wilson, E. O. 1963. "An Equilibrium Theory of Insular Zoogeography." *Evolution* **17**: 373–387.

MacArthur, R. H., and Wilson, E. O. 1967. *The Theory of Island Biogeography*. Princeton, NJ: Princeton University Press.

Mace, G. M. 1994. "An Investigation into Methods for Categorizing the Conservation Status of Species." In Edwards, P. J., May, R. M., and Webb, N. R., eds., *Large Scale Ecology and Conservation Biology*. Oxford: Blackwell, pp. 295–314.

Mace, G. M. 1995. "Classification of Threatened Species and Its Role in Conservation Planning." In Lawton, J. H., and May, R. M., eds., *Extinction Rates*. Oxford: Oxford University Press, pp. 197–213.

Mackey, B. G., Nix, H. A., Stein, J. A., Cork, S. E., and Bullen, F. T. 1989. "Assessing the Representative of the Wet Tropics of Queensland World Heritage Property." *Biological Conservation* **50**: 279–303.

Magurran, A. E. 1988. *Ecological Diversity and Its Measurement*. Princeton, NJ: Princeton University Press.

Maisto, S. A. 2003. *Drug Use and Abuse*. Belmont, CA: Wadsworth.

Malpas, J. E. 1999. *Place and Experience: A Philosophical Topography*. Cambridge: Cambridge University Press.

Manly, B. F. J., McDonald, L. L., and Thomas, D. L. 1993. *Resource Selection by Animals: Statistical Design and Analysis for Field Studies*. London: Chapman and Hall.

Mann, C. C., and Plummer, M. L. 1995. *Noah's Choice: The Future of Endangered Species*. New York: Knopf.

Marchak, M. P. 1995. *Logging the Globe*. Montreal: McGill–Queens University Press.

Margules, C. R. 1989. "Introduction to Some Australian Developments in Conservation Evaluation." *Biological Conservation* **50**: 1–11.

Margules, C. R., Higgs, A. J., and Rafe, R. W. 1982. "Modern Biogeographic Theory: Are There Lessons for Nature Reserve Design?" *Biological Conservation* **24**: 115–128.

Margules, C. R., Nicholls, A. O., and Pressey, R. L. 1988. "Selecting Networks of Reserves to Maximize Biological Diversity." *Biological Conservation* **43**: 63–76.

Margules, C. R., and Pressey, R. L. 2000. "Systematic Conservation Planning." *Nature* **405**: 242–253.

Martin, R. D. 1993. "Primate Origins: Plugging the Gaps." *Nature* **363**: 223–234.

Mathews, F. 2001. "Deep Ecology." In Jamieson, D., ed., *A Companion to Environmental Philosophy*. Oxford: Blackwell, pp. 218–232.

May, R. M. 1973. *Stability and Complexity in Model Ecosystems*. Princeton, NJ: Princeton University Press.

May, R. M. 1975. "Island Biogeography and the Design of Wildlife Preserves." *Nature* **254**: 177–178.

May, R. M., Lawton, J. H., and Stork, N. E. 1995. "Assessing Extinction Rates." In Lawton, J. H., and May, R. M., eds., *Extinction Rates*. Oxford: Oxford University Press, pp. 1–24.

Maynard Smith, J. 1975. *The Theory of Evolution*, 3rd. ed. Harmondsworth: Penguin.

Mayr, E. 1976. *Evolution and the Diversity of Life: Selected Essays*. Cambridge, MA: Harvard University Press.

McCann, K. S. 2000. "The Diversity-Stability Debate." *Nature* **405**: 228–233.

McCully, P. 1996. *Silenced Rivers: The Ecology and Politics of Large Dams*. London: Zed Books.

McDonald, K. R. 1990. "*Rheobatachrus* Liem and *Taudactylus* Straughan and Lee (Anura: Leptodactylidae) in Eungella National Park, Queensland: Distribution and Decline." *Transactions of the Royal Society of South Australia* **114**: 187–194.

McIntosh, R. P. 1985. *The Background of Ecology: Concept and Theory*. Cambridge: Cambridge University Press.

McLellan, D. 1986. *Ideology*. Milton Keynes, UK: Open University Press.

Meffe, G. K., and Carroll, C. R. 1994. *Principles of Conservation Biology*. Sunderland, MA: Sinauer Associates.

Mill, J. S. 1962. *Utilitarianism, On Liberty, Essay on Bentham*. New York: Meridian.

Moffett, A., Dyer, J., and Sarkar, S. 2005. "Integrating Biodiversity Representation with Multiple Criteria in North-Central Namibia Using Non-Dominated Alternatives and a Modified Analytic Hierarchy Process."

Moore, G. E. 1903. *Principia Ethica*. Cambridge: Cambridge University Press.

Morris, S. C. 1995. "New Phylum from the Lobster's Lips." *Nature* **378**: 661–662.

Mrosovsky, N. 2001. "The Future of Ridley Arribadas in Orissa: From Triple Waste to Triple Win." *Kachhapa* **5**: 1–3.

Mukerjee, R. K. 1942. *Social Ecology*. London: Green and Co.

Munansinghe, M. 1993. "Environmental Economics and Sustainable Development." Washington, DC: World Bank.

Munroe, E. G. 1948. "The Geographical Distribution of Butterflies in the West Indies." Ph.D. dissertation, Cornell University.

Nabhan, G. P. 1995. "Cultural Parallax in Viewing North American Habitats." In Soulé, M. E., and Lease, G., eds., *Reinventing Nature? Responses to Postmodern Deconstruction*. Washington, DC: Island Press, pp. 87–101.

Nabhan, G. P. 1997. *Cultures of Habitat: On Nature, Culture, and Story*. Washington, DC: Counterpoint.

Nabhan, G. P., Rea, A. M., Reichardt, K. L., Mellink, E., and Futchinson, E. F. 1982. "Papago Influences on Habitat and Biotic Diversity: Quitovac Oasis Ethnoecology. *Journal of Ethnobiology* **2**: 124–143.

Naeem, S. 1998. "Species Redundancy and Ecosystem Reliability." *Conservation Biology* **12**: 39–45.

Naeem, S. 2002. "Biodiversity Equals Instability?" *Nature* **416**: 23–24.

Naess, A. 1986. "Intrinsic Value: Will the Defenders of Nature Please Rise?" In Soulé, M., ed., *Conservation Biology: The Science of Scarcity and Diversity*. Sunderland, MA: Sinauer, pp. 504–515.

Naess, A. 1995 [1973]. "The Shallow and the Deep, Long-Range Ecology Movement." In Drengson, A., and Inoue, Y., eds., *The Deep Ecology Movement: An Introductory Anthology*. Berkeley, CA: North Atlantic Books, pp. 3–9.

Naess, A., and Sessions, G. 1995 [1984]. "Platform Principles of the Deep Ecology Movement." In Drengson, A., and Inoue, Y., eds., *The Deep Ecology Movement: An Introductory Anthology*. Berkeley, CA: North Atlantic Books, pp. 49–53.

Nash, R. 1973. *Wilderness and the American Mind*, 2nd ed. New Haven, CT: Yale University Press.

Nash, R. 1982. *Wilderness and the American Mind*, 3rd ed. New Haven, CT: Yale University Press.

Nicholls, A. O. 1989. "How to Make Biological Surveys Go Further with Generalised Linear Models." *Biological Conservation* **50**: 51–75.

Nix, H. A., Faith, D. P., Hutchinson, M. F., Margules, C. R., West, J., Allison, A., Kesteven, J. L., Natera, G., Slater, W., Stein, J. L., and Walker, P. 2000. *The BioRap*

Toolbox: A National Study of Biodiversity Assessment and Planning for Papua New Guinea: Consultancy Report to the World Bank. Canberra: CSIRO Press.

Norse, E. A. 1990. *Ancient Forests of the Pacific Northwest*. Washington, DC: Island Press.

Norton, B. G. 1987. *Why Preserve Natural Variety?* Princeton, NJ: Princeton University Press.

Norton, B. G. 1991. *Toward Unity among Environmentalists*. Oxford: Oxford University Press.

Norton, B. G. 2003. *Searching for Sustainability*. New York: Cambridge University Press.

Noss, R. 1984. "Deep Ecology, Elitism and Reproduction." *Earth First!* **4** (5): 16.

Nozick, R. 1974. *Anarchy, State, and Utopia*. New York: Basic Books.

Odenbaugh, J. 2005. "Ecology." In Sarkar, S., and Pfeifer, J., eds., *The Philosophy of Science: An Encyclopedia*. New York: Routledge, in press.

Oelschlaeger, M. 1991. *The Idea of Wilderness: From Prehistory to the Age of Ecology*. New Haven, CT: Yale University Press.

O'Neill, J. 1992. "The Varieties of Intrinsic Value." *The Monist* **75**: 119–137.

O'Neill, J. 1993. *Ecology, Policy and Politics: Human Well-Being and the Natural World*. London: Routledge.

O'Neill, J. 2001. "Meta-ethics." In Jamieson, D., ed., *A Companion to Environmental Philosophy*. Malden, MA: Blackwell, pp. 163–176.

Orians, G. H., and Heerwagen, J. H. 1992. "Evolved Responses to Landscapes." In Barkow, J. H., Cosmides, L., and Tooby, J., eds., *The Adapted Mind: Evolutionary Psychology and the Generation of Culture*. Oxford: Oxford University Press, pp. 555–579.

Orzack, S. H., and Sober, E. 1993. "A Critical Assessment of Levins's *The Strategy of Model Building in Population Biology* (1966)." *Quarterly Review of Biology* **68**: 533–546.

Pandav, B., and Kar, C. S. 2000. "Reproductive Span of Olive Ridley Turtles at Gahirmatha Rookery, Orissa, India." *Marine Turtle Newsletter* **87**: 8–9.

Pearce, D. W. 1993. *Economic Value and the Natural World*. Cambridge, MA: MIT Press.

Peet, R., and Watts, M., eds. 1996. *Liberation Ecologies: Environment, Development, Social Movements*. London: Routledge.

Pence, G. E. 1998. *Who's Afraid of Human Cloning?* Oxford: Rowman and Lilltefield.

Peters, R. H. 1991. *A Critique for Ecology*. Cambridge: Cambridge University Press.

Pfisterer, A., and Schmid, B. 2002. "Diversity-Dependent Production Can Decrease Stability of Ecosystem Functioning." *Nature* **416**: 84–86.

Pianka, E. R. 2000. *Evolutionary Ecology*, 6th. ed., San Francisco: Benjamin-Cummings.

Pickett, S. T. A., Kolasa, J., and Jones, C. G. 1994. *Ecological Understanding*. San Diego, CA: Academic Press.

Pimm, S. L. 1991. *The Balance of Nature? Ecological Issues in the Conservation of Species and Communities*. Chicago: University of Chicago Press.

Pimm, S. L., and van Aarde, R. J. 2001. "African Elephants and Contraception." *Nature* **411**: 766.

Plumwood, V. 1998. "Wilderness Skepticism and Wilderness Dualism." In Callicott, J. B., and Nelson, M. P., eds., *The Great New Wilderness Debate*. Athes: University of Georgia Press, pp. 652–690.

Polasky, S., Camm, J. D., Solow, A. R., Csuti, B., White, D., and Ding, R. 2000. "Choosing Reserve Networks with Incomplete Species Information." *Biological Conservation* **94**: 1–10.

Possingham, H., Ball, I., and Andelman, S. 2000. "Mathematical Methods for Identifying Representative Reserve Networks." In Ferson, S., and Burgman, M., eds., *Quantitative Methods for Conservation Biology*. New York: Springer-Verlag, pp. 291–305.

Pounds, J. A., Fogden, M. P. L., and Campbell, J. H. 1999. "Biological Response to Climate Change on a Tropical Mountain." *Nature* **398**: 611–615.

Power, M. E., Tilman, D., Estes, J. A., Menge, B. A., Bond, W. J., Mills, L. S., Daily, G., Castilla, J. C., Lubchenco, J., and Paine, R. T. 1996. "Challenges in the Quest for Keystones." *BioScience* **46**: 609–620.

Prendergast, J. R., Quinn, R. M., and Lawton, J. H. 1999. "The Gaps between Theory and Practice in Selecting Nature Reserves." *Conservation Biology* **13**: 484–492.

Prentice, I. C., Farquhar, G. D., Fasham, M. J. R., Goulden, M. L., Heimann, M., Jaramillo, V. J., Kheshgi, H. S., Le Quere, C., Scholes, R. J., and Wallace, D. W. R. 2001. "The Carbon Cycle and Atmospheric Carbon Dioxide." In Houghton, J. T., Ding, Y., Griggs, D. J., Noguer, M., van der Linden, P. J., Dai, X., Maskell, K., and Johnson, C. A., eds., *Climate Change 2001: The Scientific Basis*. Cambridge: Cambridge University Press, pp. 183–237.

Pressey, R. L. 1994. "*Ad Hoc* Reservations: Forward or Backward Steps in Developing Representative Reserve Systems." *Conservation Biology* **8**(3): 662–668.

Pressey, R. L., and Cowling, R. M. 2000. "Reserve Selection Algorithms and the Real World." *Conservation Biology* **15**: 275–277.

Preston, F. W. 1962a. "The Canonical Distribution of Commonness and Rarity: Part I." *Ecology* **43**: 185–215.

Preston, F. W. 1962b. "The Canonical Distribution of Commonness and Rarity: Part II." *Ecology* **43**: 410–432.

Primack, R. B. 1993. *Essentials of Conservation Biology*. Sunderland, MA: Sinauer.

Puccia, C. J., and Levins, R. 1985. *Qualitative Modeling of Complex Systems*. Cambridge, MA: Harvard University Press.

Pyne, S. J. 1982. *Fire in America: A Cultural History of Wildland and Rural Fire*. Princeton, NJ: Princeton University Press.

Rabinowitz, A. 2001. *Beyond the Last Village: A Journey of Discovery in Asia's Forbidden Wilderness*. Washington, DC: Island Press.

Rabinowitz, D., Cairns, S., and Dillon, T. 1986. "Seven Forms of Rarity and Their Frequency in the Flora of the British Isles." In Soulé, M. E., ed., *Conservation Biology: The Science of Scarcity and Diversity*. Sunderland, MA: Sinauer, pp. 182–204.

Raup, D. M. 1976. "Species Diversity on the Phanerozoic: A Tabulation." *Paleobiology* **4**: 1–15.

Raup, D. M. 1978. "Cohort Analysis of Generic Survivorship." *Paleobiology* **4**: 1–15.

Raup, D. M. 1986. "Biological Extinction in Earth History." *Science* **231**: 1528–1533.

Raup, D. M. 1991. *Extinction: Bad Genes or Bad Luck?* New York: Norton.

Raup, D. M. 1992. "Large-Body Impact and Extinction in the Phanerozoic." *Paleobiology* **18**: 80–88.

Raup, D. M., and Sepkoski, J. J. 1984. "Periodicity of Extinctions in the Geologic Past." *Proceedings of the National Academy of Sciences (USA)* **81**: 801–805.

Real, L. A., and Brown, J. H., eds. 1991. *Foundations of Ecology: Classic Papers with Commentaries*. Chicago: University of Chicago Press.

Rebelo, A. G., and Siegfried, W. R. 1990. "Protection of Fynbos Vegetation: Ideal and Real-World Options." *Biological Conservation* **54**: 15–31.

Redford, K., 1990. "The Ecologically Noble Savage." *Orion Nature Quarterly* **9**(3): 24–29.

Redford, K., and Sanjayan, M. A. 2003. "Retiring Cassandra." *Conservation Biology* **17**: 1473–1474.

Regan, T. 1983. *The Case for Animal Rights*. Berkeley: University of California Press.

Regan, T. 1992. "Does Environmental Ethics Rest on a Mistake?" *Monist* **75**: 161–183.

ReVelle, C. S., Williams, J. C., and Boland, J. J. 2002. "Counterpart Models in Facility Location Science and Reserve Selection Science." *Environmental Modeling and Assessment* **7**: 71–80.

Revkin, A. 1990. *The Burning Season*. Boston: Houghton Mifflin.

Robert, C. P. 2001. *The Bayesian Choice*, 2nd ed. Berlin: Springer.

Robert, J. S. 2004. *Embryology, Epigenesis and Evolution: Taking Development Seriously*. New York: Cambridge University Press.

Robichaud, W. G. 1998. "Physical and Behavioral Description of a Captive Saola, *Pseudoryx nghetinhensis*." *Journal of Mammalogy* **79**: 394–405.

Rolston, H., III. 1988. *Environmental Ethics: Duties to and Values in the Natural World*. Philadelphia: Temple University Press.

Rolston, H., III. 1991. "The Wilderness Idea Reaffirmed." *Environmental Professional* **13**: 370–377.

Rosenzweig, M. L. 2005. "Applying Species-Area Relationships to the Conservation of Species Diversity." In Lomolino, M. V., and Heaney, L. R., eds., *Frontiers of Biogeography: New Directions in the Geography of Nature*. Sunderland, MA: Sinauer Associates, pp. 325–344.

Running, S. W., Nemani, R. R., Peterson, D. L., Band, L. E., Potts, D. F., Pierce, L. L., and Spanner, M. A. 1989. "Mapping Regional Forest Evapotranspiration and Photosynthesis by Coupling Satellite Data with Ecosystem Simulation." *Ecology* **70**: 1090–1101.

Ryan, M. G. 1991. "Effects of Climate Change on Plant Respiration." *Ecological Applications* **1**: 157–167.

Saarinen, E., ed. 1982. *Conceptual Issues in Ecology*. Dordrecht: Reidel.

Saaty, T. L. 1980. *The Analytic Hierarchy Process*. New York: McGraw-Hill.

Sader, S. A., and Joyce, A. T. 1988. "Deforestation Rates and Trends in Costa Rica." *Biotropica* **20**: 11–19.

Sagoff, M. 1984. "Animal Liberation and Environmental Ethics: Bad Marriage, Quick Divorce." *Osgood Hall Law Journal* **22**: 297–307.

Sagoff, M. 1991. "Zuckerman's Dilemma: A Plea for Environmental Ethics." *Hastings Center Report* **21**: 32–40.

Salleh, A. K. 1984. "Deeper than Deep Ecology: The Eco-Feminist Connection." *Environmental Ethics* **6**: 339–345.

Sarakinos, H., Nicholls, A. O., Tubert, A., Aggarwal, A., Margules, C. R., and Sarkar, S. 2001. "Area Prioritization for Biodiversity Conservation in Québec on the Basis of Species Distributions: A Preliminary Analysis." *Biodiversity and Conservation* **10**: 1419–1472.

Sarkar, S. 1996. "Ecological Theory and Anuran Declines." *BioScience* **46**: 199–207.

Sarkar, S. 1998a. *Genetics and Reductionism*. New York: Cambridge University Press.

Sarkar, S. 1998b. "Restoring Wilderness or Reclaiming Forests?" *Terra Nova* **3**(3): 35–52.

Sarkar, S. 1998c. "Wallace's Belated Revival." *Journal of Biosciences* **23**: 3–7.

Sarkar, S. 1999. "Wilderness Preservation and Biodiversity Conservation – Keeping Divergent Goals Distinct." *BioScience* **49**: 405–412.

Sarkar, S. 2002. "Defining 'Biodiversity': Assessing Biodiversity." *Monist* **85**: 131–155.

Sarkar, S. 2003. "Conservation Area Networks." *Conservation and Society* **1**: v–vii.

Sarkar, S. 2004. "Conservation Biology." In Zalta, E. N., ed., *The Stanford Encyclopedia of Philosophy* (Summer 2004 edition). <http://plato.stanford.edu/archives/sum2004/entries/conservation-biology/>.

Sarkar, S., Aggarwal, A., Garson, J., Margules, C. R., and Zeidler, J. 2002. "Place Prioritization for Biodiversity Content." *Journal of Biosciences* **27** (S2): 339–346.

Sarkar, S., and Garson, J. 2004. "Multiple Criterion Synchronizations for Conservation Area Network Design: The Use of Non-Dominated Alternative Sets." *Conservation and Society* **2**: 433–448.

Sarkar, S., Justus, J., Fuller, T., Kelley, C., Garson, J., and Mayfield, M. 2005. "Effectiveness of Environmental Surrogates for the Selection of Conservation Area Networks." *Conservation Biology*, in press.

Sarkar, S., and Margules, C. R. 2002. "Operationalizing Biodiversity for Conservation Planning." *Journal of Biosciences*: **27** (S2): 299–308.

Sarkar, S., Pappas, C., Garson, J., Aggarwal, A., and Cameron, S. 2004. "Place Prioritization for Biodiversity Conservation Using Probabilistic Surrogate Distribution Data." *Diversity and Distributions* **10**: 125–133.

Sarkar, S., Parker, N. C., Garson, J., Aggarwal, A., and Haskell, S. 2000. "Place Prioritization for Texas Using GAP Data: The Use of Biodiversity and Environmental Surrogates within Socioeconomic Constraints." *Gap Analysis Program Bulletin* **9**: 48–50.

Sartre, J. P. 1949. *Nausea*. Norfolk, CT: New Directions.

Sauer, J. D. 1969. "Oceanic Islands and Biogeographical Theory: A Review." *Geographical Review* **59**: 582–593.

Savidge, J. A. 1987. "Extinction of an Island Avifauna by an Introduced Snake." *Ecology* **68**: 660–668.

Schaller, G., and Vrba, E. S. 1996. "Description of the Giant Muntjac *Megamuntiacus vuquangensis* in Laos." *Journal of Mammalogy* **77**: 675–683.

Schultes, R. E. 1976. *Hallucinogenic Plants*. New York: Golden Press.

Schweitzer, A. 1976. *Civilization and Ethics*. Englewood Cliffs, NJ: Prentice-Hall.

Scott, J. M., Heglund, P. J., Morrison, M. L., Haufler, J. B., Raphael, M. G., Wall, W. A., and Samson, F. B., eds. 2002. *Predicting Species Occurrences: Issues of Accuracy and Scale*. Washington, DC: Island Press.

Scudo, F. M., and Ziegler, J. R., eds. 1978. *The Golden Age of Theoretical Ecology, 1923–1940*. Berlin: Springer-Verlag.

Sellers, P. J., Dickinson, R. E., Randall, D. A., Betts, A. K., Hall, F. G., Berry, J. A., Collatz, G. J., Denning, A. S., Mooney, H. A., Nobre, C. A., Sato, N., Field, C. B., and Henderson-Sellers, A. 1997. "Modeling the Exchanges of Energy, Water, and Carbon between the Continents and the Atmosphere." *Science* **275**: 502–509.

Shafer, C. L. 1999. "National Park and Reserve Planning to Protect Biological Diversity: Some Basic Elements." *Landscape and Urban Planning* **44**: 123–153.

Shaffer, M. L. 1978. "Determining Minimum Viable Population Sizes: A Case Study of the Grizzly Bear." Ph.D. dissertation, Duke University.

Shaffer, M. L. 1981. "Minimum Population Sizes for Species Conservation." *BioScience* **31**: 131–134.

Shaffer, M. L. 1987. "Minimum Viable Populations: Coping with Uncertainty." In Soulé, M. E., ed., *Viable Populations for Conservation*. Cambridge: Cambridge University Press, pp. 69–86.

Shannon, C. E. 1948. "A Mathematical Theory of Communication." *Bell System Technical Journal* **27**: 379–423, 623–656.

Shiva, V. 1991. "Biodiversity, Biotechnology and Profits." In Shiva, V., Anderson, P., Schücking, H., Gray, A. Lohmann, L., and Cooper, D., *Biodiversity: Social and Ecological Perspectives*. London: Zed Books, pp. 43–58.

Shrader-Frechette, K. S. 1990. "Island Biogeography, Species-Area Curves, and Statistical Errors: Applied Biology and Scientific Rationality." In Fine, A., Forbes, M., and Wessels, L., eds., *PSA 1990: Proceedings of the 1990 Biennial Meeting of the Philosophy of Science Association*, vol. 1. East Lansing, MI: Philosophy of Science Association, pp. 447–456.

Shrader-Frechette, K. S. 1991. *Risk and Rationality*. Berkeley: University of California Press.

Shrader-Frechette, K. S. 2002. *Environmental Justice: Creating Equality, Reclaiming Democracy*. New York: Oxford University Press.

Shrader-Frechette, K. S., and McCoy, E. D. 1993. *Method in Ecology: Strategies for Conservation*. Cambridge: Cambridge University Press.

Shugart, H. H. 1984. *A Theory of Forest Dynamics: The Ecological Implications of Forest Succession Models*. New York: Springer-Verlag.

Shugart, H. H., Smith, T. M., and Post, W. M. 1992. "The Potential Application of Individual-Based Simulation Models for Assessing the Effects of Global Change." *Annual Review of Ecology and Systematics* **23**: 15–38.

Simberloff, D. S., and Abele, L. G. 1976a. "Island Biogeography Theory and Conservation Practice." *Science* **191**: 285–286.

Simberloff, D. S., and Abele, L. G. 1976b. "Island Biogeography Theory and Conservation: Strategy and Limitations." *Science* **193**: 1032.

Singer, P. 1975. *Animal Liberation*. New York: New York Review.

Singer, P. 1979. *Practical Ethics*. Cambridge: Cambridge University Press.

Singer, P. 2001. "Animals." In Jamieson, D., ed., *A Companion to Environmental Philosophy*. Oxford: Blackwell, pp. 416–425.

Siskind, J. 1973. *To Hunt in the Morning*. New York: Oxford University Press.

Smith, F. D. M., May, R. M., Pellew, R., Johnson, T. H., and Walker, K. S. 1993a. "Estimating Extinction Rates." *Nature* **364**: 494–496.

Smith, F. D. M., May, R. M., Pellew, R., Johnson, T. H., and Walker, K. S. 1993b. "How Much Do We Know about the Current Extinction Rate?" *Trends in Ecology and Evolution* **8**: 375–378.

Sober, E. 1986. "Philosophical Problems for Environmentalism." In Norton, B. G., ed., *The Preservation of Species: The Value of Biological Diversity*. Princeton, NJ: Princeton University Press, pp. 173–194.

Sokal, R., and Crovello, T. 1970. "The Biological Species Concept: A Critical Evaluation." *American Naturalist* **104**: 127–153.

Soper, K. 1995. *What Is Nature? Culture, Politics, and the Non-Human*. Oxford: Blackwell.

Soulé, M. E. 1985. "What Is Conservation Biology?" *BioScience* **35**: 727–734.

Soulé, M. E., 1987. "History of the Society for Conservation Biology: How and Why We Got Here." *Conservation Biology* **1**: 4–5.

Soulé, M. E., and Kohm, K. 1989. *Research Priorities in Conservation Biology*. Washington, DC: Island Press.

Soulé, M. E., and Orians, G. H., eds. 2001. *Conservation Biology: Research Priorities for the Next Decade*. Washington, DC: Island Press.

Soulé, M. E., and Sanjayan, M. A. 1998. "Conservation Targets: Do They Help?" *Science* **279**: 2060–2061.

Soulé, M. E., and Simberloff, D. S. 1986. "What Do Genetics and Ecology Tell Us about the Design of Nature Reserves?" *Biological Conservation* **35**: 19–40.

Soulé, M. E., and Terborgh, J., eds. 1999. *Continental Conservation: Scientific Foundations of Regional Reserve Networks*. Washington DC: Island Press.

Soulé, M. E., Wilcox, B. A., and Holtby, C. 1979. "Benign Neglect: A Model of Faunal Collapse in the Game Reserves of East Africa." *Biological Conservation* **15**: 259–272.

Spence, M. D. 1999. *Dispossessing the Wilderness: Indian Removal and the Making of the National Parks*. New York: Oxford University Press.

Spencer, M. 1997. "The Effects of Habitat Size and Energy on Food Web Structure: An Individual-Based Cellular Automata Model." *Ecological Modelling* **94**: 299–316.

Stevens, S., ed., 1997. *Conservation through Cultural Survival*. Washington, DC: Island Press.

Stockwell, D., and Peters, D. 1999. "The GARP Modelling System: Problems and Solutions to Automated Spatial Prediction." *International Journal of Geographical Information Science* **13**: 143–148.

Stone, C. D. 1974. *Should Trees Have Standing? Toward Legal Rights for Natural Objects*. Los Altos, CA: William Kaufmann.

Stone, C. D. 1985. "*Should Trees Have Standing?* Revisited: How Far Will Law and Morals Reach? A Pluralist Perspective." *Southern California Law Review* **59**: 1.

Sugihara, G. 1981. "$S = CA^z$, $z \approx 1/4$: A Reply to Connor and McCoy." *American Naturalist* **117**: 790–793.

Sukumar, R. 1989. *The Asian Elephant: Ecology and Management*. Cambridge: Cambridge University Press.

Sukumar, R. 1994. *Elephant Days and Nights: Ten Years with the Asian Elephant*. New Delhi: Oxford University Press.

Sulloway, F. J. 1982. "Darwin and His Finches: The Evolution of a Legend." *Journal of the History of Biology* **15**: 1–53.

Sylvan, R., and Bennett, D. 1994. *The Greening of Ethics: From Human Chauvinism to Deep-Green Theory*. Cambridge, MA: Whitehorse Press.
Takacs, D. 1996. *The Idea of Biodiversity: Philosophies of Paradise*. Baltimore: Johns Hopkins University Press.
Tansley, A. G. 1935. "The Use and Abuse of Vegetational Concepts and Terms." *Ecology* **16**: 284–307.
Taylor, B. L., Wade, P. R., Stehn, R. A., and Cochrane, J. F. 1996. "A Bayesian Approach to Classification Criteria for Spectacled Eiders." *Ecological Applications* **6**: 1077–1089.
Taylor, P. W. 1986. *Respect for Nature: A Theory of Environmental Ethics*. Princeton, NJ: Princeton University Press.
Terborgh, J. 1975. "Faunal Equilibria and the Design of Wildlife Preserves." In Golley, F., and Medina, E., eds., *Tropical Ecological Systems: Trends in Terrestrial and Aquatic Research*. New York: Springer, pp. 369–380.
Terborgh, J. 1976. "Island Biogeography and Conservation: Strategy and Limitations." *Science* **193**: 1029–1030.
Thompson, J. 1995. "Aesthetics and the Value of Nature." *Environmental Ethics* **17**: 291–305.
Tilman, D. 1999. "The Ecological Consequences of Biodiversity: A Search for General Principles." *Ecology* **80**: 1455–1474.
Tomlinson, R. F. 1988. "The Impact of the Transition from Analogue to Digital Cartographic Representation." *American Cartographer* **15**: 249–261.
Tyler, M. J., ed. 1983. *The Gastric Brooding Frog*. Canberra: Croom Helm.
Tyler, M. J., and Davies, M. 1985. "The Gastric Brooding Frog *Rheobatachrus silus*." In Grigg, G., Shine, R., and Ehmann, H., eds., *Biology of Australian Frogs and Reptiles*. Chipping Norton, New South Wales, Australia: Surrey Beatty and Sons, pp. 469–470.
Underhill, L. G. 1994. "Optimal and Suboptimal Reserve Selection Algorithms." *Biological Conservation* **70**: 85–87.
van Gelder, S. and Rysavy, T. 1998. "Dividing the Sierra Club." *Yes!* (**6**): 6.
Vane-Wright, R. I., Humphries, C. J., and Williams, P. H. 1991. "What to Protect? – Systematics and the Agony of Choice." *Biological Conservation* **55**: 235–254.
Varner, G. E. 1987. "Do Species Have Standing?" *Environmental Ethics* **9**: 57–72.
Varner, G. E. 1998. *In Nature's Interests? Interests, Animal Rights, and Environmental Ethics*. New York: Oxford University Press.
Varner, G.E. 2001. "Sentientism." In Jamieson, D., ed., *A Companion to Environmental Philosophy*. Malden, MA: Blackwell, pp. 192–203.
Vijayan, V. S. 1987. *Keoladeo National Park Ecology Study*. Bombay: Bombay Natural History Society.
Vitousek, P. M. 1994. "Beyond Global Warming: Ecology and Global Change." *Ecology* **75**: 1861–1876.
Volterra, V. [1927] 1978. "Variations and Fluctuations in the Numbers of Coexisting Animal Species." In Scudo, F. M., and Ziegler, J. R., eds., *The Golden Age of Theoretical Ecology: 1923–1940*. Berlin: Springer-Verlag, pp. 65–236.
von Wright, G. H. 1963. *The Varieties of Goodness*. London: Routledge and Kegan Paul.

Voss, J., and Sarkar, S. 2003. "Depictions as Surrogates for Places: From Wallace's Biogeography to Koch's Dioramas." *Philosophy & Geography* **6**: 60–81.

Wade, P. R. 1994. "Abundance and Population Dynamics of Two Eastern Tropical Pacific Dolphins, *Stenella attenuata* and *Stenella attenuata orientalis*." Ph. D. dissertation, University of California, San Diego.

Wade, P. R. 2000. "Bayesian Methods in Conservation Biology." *Conservation Biology* **14**: 1308–1316.

Wade, P. R. 2002. "Bayesian Population Viability Analysis." In Beissinger, S. R., and McCullough, D. R., eds., *Population Viability Analysis*. Chicago: University of Chicago Press, pp. 213–238.

Wallace, A. R. 1855. "On the Law Which Has Regulated the Introduction of New Species." *Annals and Magazine of Natural History* **16**: 184–196.

Wallace, A. R. 1858. "On the Tendency of Varieties to Depart Indefinitely from the Original Type." *Journal of the Proceedings of the Linnaean Society. Zoology* **3**: 53–62.

Wenz, P. S. 1988. *Environmental Justice*. Albany: State University of New York Press.

Western, D., and Ssemakula, J. 1981. "The Fauna of Savannah Ecosystems: Ecological Islands or Faunal Enclaves?" *African Journal of Ecology* **19**: 7–19.

Whitcomb, R. F., Lynch, J. F., and Opler, P. A. 1976. "Island Biogeography and Conservation: Strategy and Limitations." *Science* **193**: 1030–1032.

White, L. 1967. "The Historical Roots of Our Ecological Crisis." *Science* **155**: 1203–1207.

Whittaker, R. H. 1972. "Evolution and the Measurement of Species Diversity." *Taxon* **21**: 213–251.

Whittaker, R. H. 1975. *Communities and Ecosystems*. New York: Macmillan.

Whittle, P. 2000. *Probability via Expectation*, 4th. ed., New York: Springer.

Williams, B. K. 1997. "Approaches to the Management of Waterfowl under Uncertainly." *Wildlife Society Bulletin* **25**: 714–720.

Williams, B. K. 2001. "Uncertainty, Learning, and the Optimal Management of Wildlife." *Environmental and Ecological Statistics* **8**: 269–288.

Williams, P., and Humphries, C. J. 1994. "Biodiversity, Taxonomic Relatedness, and Endemism in Conservation." In Forey, P. L., Humphries, C. J., and Vane-Wright, R. I., eds., *Systematics and Conservation Evaluation*. Oxford: Clarendon Press, pp. 269–287.

Wilson, E. O. 1988a. *Biophilia: The Human Bond with Other Species*. Cambridge, MA: Harvard University Press.

Wilson, E. O. Ed. 1988b. *BioDiversity*. Washington, DC: National Academy Press.

Wilson, E. O., and Willis, E. O. 1975. "Applied Biogeography." In Cody, M. L., and Diamond, J. M., eds., *Ecology and the Evolution of Communities*. Cambridge, MA: Harvard University Press, pp. 522–534.

Wimsatt, W. C. 1987. "False Models as Means to Truer Theories." In Nitecki, M., and Hoffman, A., eds., *Natural Modes in Biology*. Oxford: Oxford University Press, pp. 23–55.

Wolf, E. 1972. "Land Ownership and Political Ecology." *Anthropological Quarterly* **45**: 201–205.

Wood, P. M. 2000. *Biodiversity and Democracy*. Vancouver: University of British Columbia Press.

Woods, M. 2001. "Wilderness." In Jamieson, D., ed., *A Companion to Environmental Philosophy*. Malden, MA: Blackwell, pp. 349–361.

Worster, D. 1994. *Nature's Economy: A History of Ecological Ideas*. Cambridge: Cambridge University Press.

Zeiler, M. 1997. *Inside ARC/INFO*. Santa Fe, NM: OnWord Press.

Zimmerman, B. L., and Bierregard, R. O. 1986. "Relevance of the Equilibrium Theory of Island Biogeography and Species-Area Relations to Conservation with a Case from Amazonia." *Journal of Biogeography* **13**: 133–143.

Index